THE IDEA AGENT

THE HANDBOOK ON CREATIVE PROCESSES

SECOND EDITION

Jonas Michanek & Andréas Breiler

Routledge
Taylor & Francis Group

NEW YORK AND LONDON

Second edition published 2014
by Routledge
711 Third Avenue, New York, NY 10017

Simultaneously published in the UK
by Routledge
2 Park Square, Milton Park, Abingdon, Oxon OX14 4RN

Routledge is an imprint of the Taylor & Francis Group, an informa business

First published by Arx Förlag 2012
Translated by Eddie Storey
© Photo of authors Jens Lennartsson
Graphic Design insert Fredrik Svensson

Library of Congress Cataloging in Publication Data
Michanek, Jonas, 1974-
[Idéagenten. English]
The idea agent : the handbook on creative processes / Jonas Michanek and Andréas
Breiler ; [Translated by Eddie Storey]. -- Second edition.
pages cm
"Authorised translation from the Swedish language edition published by Arx Förlag."
Includes bibliographical references and index.
1. Creative ability in business. 2. Creative ability. 3. Knowledge management. 4.
Technological innovations. I. Breiler, Andréas, 1973- II. Michanek, Jonas, 1974-
Idéagenten. Translation of: III. Title.
HD53.M53513 2013
658.4'094--dc23
2012045695
ISBN: 978-0-415-82414-9 (pbk)
ISBN: 978-0-203-54914-8 (ebk)

Reviews and endorsements

"*The Idea Agent* is an educational pearl; it gives insight into the authors many years of experience with managing innovation and creative processes in a way that puts this book in a league of its own. There is only one thing to say after having read this book: Welcome to the world of creativity, ideas and opportunity. *The Idea Agent* is a book that makes you both wise and happy."

Uffe Elbæk
CULTURAL MINISTER OF DENMARK AND FOUNDER OF THE KAOSPILOT UNIVERSITY

"Without Creativity there is no Innovation, but without Innovation creative ideas wither and die. This excellent book describes the role and vital contribution of *The Idea Agent*, the sympathetic leader who guides his team through the tangled and difficult process of marshalling creative discovery and transforming it into to innovative products. The second half of the book is a comprehensive and invaluable description of the many tools that the Idea Agent can use to facilitate this process, but the overwhelming message is that the process must be managed, that one must never lose sight of the need which is the necessary starting point for all creative endeavor, and that it is the Ideal Agent who is key to success."

Colin Alexander
CHAIRMAN, EU INCUBATOR FORUM, AND HEAD OF CONSULTANCY, OXFORD INNOVATION LTD

"Most companies don't have a shortage of ideas, but a shortage of the right ones. How to best find and develop ideas is the very foundation for all innovation. *The Idea Agent* hits the bull's eye with a practical handbook you want to read again and again."

Jørgen Thorball
DIRECTOR OF NOVOZYMES BIOTECH BUSINESS DEVELOPMENT AND CO-FOUNDER OF THE DANISH INNOVATION COUNCIL

"The growing importance of innovation can almost be measured by the explosion in the number of books published on the subject. But despite the acknowledged need to make innovation more accessible, so many books seem to make it complex or specialize in one area of innovation. In this sense, *The Idea Agent* is a great achievement. It is great to see all the key aspects of idea management brought together so practically in one place, in a book that doesn't over claim and is a pleasure to read."

Paul Sigsworth
CREATIVITY DEVELOPMENT, NESTLÉ ROWNTREE

"I congratulate the writers and the forthcoming readers on a book that is truly hands-on. I am impressed by all the practical process design suggestions that are in the book. *The Idea Agent* is a practical guide on how to reach the window of opportunity in the age of creativity."

Leif Edvinsson
PROFESSOR, INTELLECTUAL CAPITAL, UNIVERSITY OF LUND — SCHOOL OF ECONOMICS AND GLOBAL BRAIN OF THE YEAR 1998

"*The Idea Agent* is an excellent contribution in increasing the understanding and knowledge of the processes that lay the foundation for successful idea management. The idea process is often poorly conducted in many companies because of the lack of knowledge. This book can fill some of the knowledge gaps that exist by presenting many practical methods and techniques which are the basis for an effective use of the intrinsic creativity of your staff."

Sven Andrén
GLOBAL PROCESS DRIVER, IDEA MANAGEMENT, TETRA PAK

"Creativity and innovation plays an ever increasing part in our society, but unfortunately we are often not equipped enough to facilitate idea generation. *The Idea Agent* addresses this problem and gives well structured guidance into the idea process and all of its phases. The book is must for the beginner and a welcome inspiration for the expert, and it does us all a favor by pointing out the importance of structured creativity!"

Michael Thomsen
MANAGING DIRECTOR R&D, INTERACTIVE INSTITUTE AND FORMER MANAGING DIRECTOR R&D, LEGO MEDIA INTERNATIONAL

CONTENTS

PREFACE   **9**

KJELL A. NORDSTRÖM ——————————————————— 9

AUTHORS' PREFACE ——————————————————— 11

1. THE AGE OF CREATIVITY   **17**

SIX TRENDS THAT ARE CHANGING THE WAY WE WORK ——————— 20

2. IDEAS, CREATIVITY AND INNOVATION   **25**

WHAT IS AN IDEA? ——————————————————— 28

WHAT IS CREATIVITY? ——————————————————— 28

WHAT IS INNOVATION? ——————————————————— 29

THE IDEA PROCESS ——————————————————— 33

IDEA PROCESSES IN DIFFERENT BUSINESS AREAS ——————— 35

3. MANAGING CREATIVE PROCESSES   **37**

ACTIVE LEADERSHIP IN CREATIVE PROCESSES ——————— 39

THE THREE ROLES OF THE IDEA AGENT ——————————— 40

THE CREATIVE PROCESS ——————————————————— 40

4. NEED   **49**

DEFINING THE FOCUS AREA ——————————————— 54

5. IDEA GENERATION   **57**

THE RULES OF CREATIVE CHAOS ——————————————— 59

THE PATH TO CORRECT IDEA GENERATION METHODS ——————— 63

6. SCREENING AND DEVELOPMENT 67

SCREENING AND DEVELOPMENT – STEP-BY-STEP ———————————— 72

7. ENRICHING AND CONCEPTUALIZING 81

DIFFERENT TYPES OF ENRICHMENT ———————————— 83

8. RESULTS: AND WHAT NEXT? 93

WHY DOES NOBODY APPRECIATE MY BRILLIANT IDEA? ———————— 95

METHODS 99

NEED ORIENTATION TOOLS 101

5 WHYS? ———————————————————————— 101

PRIORITIZATION ANALYSIS ———————————————— 103

RADICAL HYPOTHESES ———————————————————— 105

THE RELATION MATRIX ———————————————————— 108

IDEA GENERATION TOOLS 111

RANDOM WORD ASSOCIATION ———————————————— 111

FORCED COMBINATIONS ———————————————————— 114

NEGATIVE IDEA GENERATION ———————————————— 117

WHAT IF? ———————————————————————— 120

THE DREAM TRIP ———————————————————————— 122

DAY PARTING ———————————————————————— 124

VISUAL CONFETTI ———————————————————————— 126

HEADLINE MANIA ———————————————————————— 128

6-3-5 ———————————————————————————— 130

FISHING STORIES ———————————————————————— 132

IDEA PROPPING ———————————————————— 134

MERLIN ————————————————————————— 136

YOUR CREATIVE IDOL ————————————————— 138

THE RELAY BATON —————————————————— 140

ZOOMING OUT ————————————————————— 142

THE LOTUS BLOSSOM ————————————————— 144

TRENDSTORMING ———————————————————— 146

UPSIDE DOWN —————————————————————— 148

TRIZ —————————————————————————— 150

SCREENING AND DEVELOPMENT TOOLS   153

THREE DOTS ———————————————————————— 153

TOP 10 ——————————————————————————— 155

THE FOUR-FIELD MATRIX —————————————————— 157

THE SPIDER WEB DIAGRAM —————————————————— 160

KESSELRING ————————————————————————— 162

INTRODUCTION AND ENERGY TOOLS   167

CIRCLE OF PORTRAITS ————————————————————— 167

HERE'S MY KEYRING ——————————————————————— 169

"ALLOW ME TO INTRODUCE…" ——————————————————— 170

BODYCOUNT —————————————————————————— 172

HOME ALONE ———————————————————————————— 173

STORM WARNING ——————————————————————————— 174

THE ROLLER COASTER ——————————————————————— 175

CONCENTRATION TO 20 —————————————————————— 176

THE DUNCE HAT ————————————————————————— 178

THE LAST SUPPER? ——————————————————————— 180

TEN THOUSAND RED ROSES ———————————————————— 182

APPENDIX 185

A. SOURCES OF INSPIRATION —————————————————— 187

B. GLOSSARY ——————————————————————————— 190

C. INDEX QUICK GUIDE TO IDEA GENERATION METHODS —————— 192

D. MODEL SCHEDULE FOR A DAY-LONG WORKSHOP ——————— 193

E. LAZY MAN'S GUIDE TO CREATIVE PROCESS DESIGN —————— 195

PREFACE TO THE SECOND EDITED EDITION

Can you buy ideas on the commodity exchanges?

If you could buy bad ideas like a commodity they would have a market price and hence a low price. Since it is said that there is a lack of good ideas, they would be expensive to begin with. If we had learned how to grow great ideas, the market would literally drown in useful suggestions. This also means good ideas would become relatively cheaper. That is how capitalism works.

With more great ideas, the world will thrive and businesses prosper. The big question is how to create the climate in which ideas are emerging and where entrepreneurs get them implemented.

The finest thing to be these days is a change agent, entrepreneur, rebel and innovator. We all know why ideas are needed but not so much about how they occur. Our contribution in the book, *Funky Business*, was to argue that it has to do with talent. I hope we have managed to describe that talent is something much bigger than competence. Those who have talent seem to have access to more ideas than their sisters and brothers. The talents are loved by their customers and clients, but feared and sometimes hated by their colleagues. The talents seem all so secure in their talent and they seem to be able to generate brilliant ideas, almost to order. Is it possible to teach yourself to become a talent, you have to wonder. Where do you find inspiration and how do you develop yourself and become an idea agent?

He who listens to advice, is wise, says a Swedish proverb. And there's no shortage of advice for entrepreneurs and innovators – but maybe the tips are too many: "Take more risks! Make a business plan! Give yourself time to do nothing! Expect the unexpected! Meet the exciting people! etc."

The question is how much it helps to get all this advice. There is an English word often interpreted incorrectly: serendipity. That does not mean luck or fate, as many dictionaries claim, but the fortunate circumstances that seem to fall upon those who have an open mind and are actively seeking ideas and solutions. Very often those who seem to be struck by serendipity, say they are just lucky.

When the Swedish world cup skiing legend, Ingemark Stenmark, takes on the issue, he says: "It is strange, but it seems that the more I practice the luckier I get."

My advice is: Make yourself lucky. Start practicing.

Kjell A. Nordström
PhD. Writer. Speaker.
(Ed. – and international business guru).

AUTHORS' PREFACE

The goal of *The Idea Agent* is to cultivate an understanding of a new world empowered by innovation, as well as to emphasize the importance and benefits of methodical idea management. Innovation may sit high on the agenda of most modern-day organizations but it is rarely defined in distinct processes. The great risk in this is that ideas will be managed in an uninformed or unstructured way as a result, which can have drastic consequences for the future development of an organization. *The Idea Agent* is a handbook of practical approaches to creative processes that will equip you for success in developing your organization – tangible, structured approaches known as idea management.

We believe that idea management ought to be as accepted and integral a part of work processes as strategy, planning, logistics and business intelligence. Since we cannot always count on ideas coming to us out of the blue, so to speak, we need to ensure as individuals, organizations and nations that the commodity of ideas as a primary ingredient is manufactured and refined as professionally as iron ore, timber and food. All project leaders, film producers and business development managers supervising an on-going project are familiar with a feeling of uncertainty over the quality of the project outcome. If only a little more time had been invested in defining needs and developing ideas into optimal concepts, results might have been twice as good. Instead, the project takes longer to complete and results are less satisfying.

Fig. A

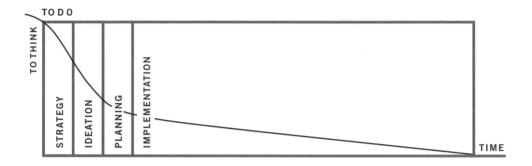

Fig. B

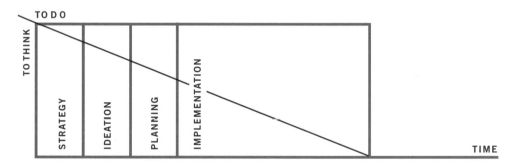

An important element in the new age of creativity is the acceptance of idea creation as an integral part of work processes. This is illustrated in the first of our models above (fig. A), which depicts in a simple way how current processes are frequently structured. The strategy, idea and planning phases should be executed quickly to reach the implementation phase as soon as possible – after all, this is actually when one can really achieve something! In reality, the entire process can usually be made much shorter and end-product quality much better if one has the foresight to allocate more time to the initial, thinking phases (fig. B).

And although we are now living in the 21st century, the Protestant work ethic still impacts on our everyday lives. In other words, our work is what characterizes us, and work is defined as effort, implementation and accomplishment. It's the seemingly physically demanding work that counts. But conceiving and creating is still not thought of as "proper" work, despite the fact that it is in developing things that the future of entire nations is defined. And if we still have to do a little creative thinking now and then, in our digital world this is usually done in front of a computer in our office. Then we're considered to be working productively although we're "only" thinking. But are we still not being fooled into following in Luther's footsteps? How often do we actually create anything substantial sitting at our computers? And don't most people claim that they think best in the shower, the bathroom or just before they fall asleep – and not in front of their computer screens? Would it not be a good idea then to take a shower at work sometimes or an occasional power nap on the couch in the coffee room? We certainly think so! After all, many of us are paid to produce our best ideas during work hours.

In brief, *The Idea Agent* is aimed at the managers of the future who are looking to play an active role in developing creative processes in all types of businesses. After reading this book and applying some of the methods it features, you'll be well on the way to becoming a professional Idea Agent. An Idea Agent is someone who can create an environment that will optimize development processes, and who keeps distinct methods up his or her figurative sleeve that will solve commonplace or future-oriented issues. Whether you're a project leader in the knowledge-intensive business sector, a product development manager in a major, product-oriented manufacturing industry, a research team leader in the biotech field or a marketing manager in a service industry, this book is for you. Anyone who has any kind of development need in their day-to-day work and takes creative processes seriously will benefit from this book. *The Idea Agent* is not a profound philosophical reflection on the essence of creativity or how to innovate on a macro level, but a guide and toolbox of methods that will enable a professional approach to the creation of ideas and concepts within a creative team. In other words, this book is an instrument to be read, noted in, thumbed through and applied in support of an Idea Agent whenever he or she is managing creative processes.

The methods contained in this book are partly developed by the authors themselves and partly elaborations of existing techniques. Identifying the origins of some of these methods is difficult but our hope is that our own ideas have both advanced and enriched them. All the methods have been tested and developed with the close and wise collaboration of clients like Tetra Pak, Absolut Vodka, Sony, Ikea, Ericsson and Carlsberg, to name a few. The book is designed for simple use in practical contexts such as establishing internal corporate structures for idea and innovation management and as a tool when coaching in creative leadership.

A huge assortment of literature exists – mostly American – that takes up creative methods and techniques, but there is also a gap in the market for a book that describes all the stages of the idea process, and one that is specific and step-by-step based. This book will hopefully fill the need and be easy to use for people both in the Western world where patent creation is high, as well as in the developed world where creativity and innovation is still lagging behind.

According to both Insead's *Global Innovation Index* 2011 and World Economic Forum's *Competitiveness Report* 2011, Sweden, Singapore and Switzerland are the most innovative countries in the world.

Our book is divided into eight chapters that encompass the areas that we believe are essential to an organization's idea production. For the parts of the book that are fundamental to idea management, there are techniques in the book's methods section that can be applied in practical situations. The **first chapter – The Age of Creativity** discusses the current trends that suggest that we are entering an Age of Creativity. **Chapter two – Ideas, Creativity, and Innovation,** and **chapter three – Managing Creative Processes** examine the creative process and the role that a facilitator plays in the work of a creative team, in other words the individual that from now on we'll be calling "the Idea Agent". The methods section provides tools that will enable team bonding and generate a positive and reassuring creative climate within your team. **Chapter four – Need** discusses how to identify a need, a central issue or an opportunity for which an idea, a concept or a solution has to be developed. Failing to establish a direction or to take the end results of a process into account can be very risky. The chapter also describes practical details that will assist you in the lead up to a planning process. The **fifth chapter – Idea Generation** deals with idea generation itself. What methods should be used? Are there any special rules when it comes to cultivating creativity. This chapter discuss everything from design of the creative process to different sources of ideas. **Chapter six – Screening and Development** describes how to take a pool of ideas and shape and organize it while sifting the wheat from the chaff using relevant matrices, in other words managing the ideas to enable an overview that will identify new opportunities for developing existing ideas and creating new ones. In **chapter seven – Enriching and Conceptualizing** we begin to examine ideas in an increasingly pragmatic way with a view to conceptualizing them. Basically, this means converting hand-written notes on a Post-it and systematically developing them into a finished concept. **Chapter eight – Results: and What Next?** deals with the part of the process that looks ahead to the realization of ideas, in other words the phase in which someone takes an idea to the next level and dares to implement it. The book concludes with an **appendix** containing a list of inspirations, a glossary of essential idea management terms, an index of methods mentioned in the text, a model schedule for a creative workshop and a lazy guide to designing creative processes.

This handbook is for the most part a result of the authors' experiences while establishing and running the consultancy firm Idélaboratoriet (**www.idelaboratoriet.se**). Idélaboratoriet works to create a culture of ideas within organizations, and collaborates with companies and organizations in a range of business sectors as well as with managers and their creative teams on most levels. The idea behind Idélaboratoriet was hatched one sunny afternoon on a café terrace on Lilla Torg in Malmö, Sweden, and was mostly inspired by studies in the Scandinavian project leadership and entrepreneurship school, 'The KaosPilot University'.

The Idea Agent could never have been written without a huge amount of help from a large number of people. One significant knowledge source was Göran Hydbom, who helped in the development and continual regeneration of our approach and in laying the groundwork, particularly for the section on screening and development, and its graphic solutions. We have also learned a great deal from our collaboration with Sven Andrén at Tetra Pak R&D, who presented us with many challenges to overcome. A big thank you as well to Johan Gunnars for his practical business sense and the well constructed models that he provided for our book. We've also received invaluable proofreading and feedback from Carl Magnus Cronholm, Anders Sjöstedt and Mikkel Thagaard, and we'd like to thank Fredrik Svensson, our graphic Sancho Panza, for his many years of effort. We would like to express our gratitude to our publisher, Hanserik Tonnhelm, and our translator, Eddie Storey, for their hard work and commitment. Finally, we would like to thank our nearest and dearest for helping us to keep faith with our ideas and realize them!

Jonas Michanek & Andréas Breiler

chapter 1

THE AGE OF CREATIVITY

– or what are the signs of the times?

"Ideas control the world"

James A. Garfield, served as 20th President of the US

Look at the world around you.
What do you see?
You probably see objects, shapes, materials... you may hear music,
sound or noise... or can you taste ice cream, feel the keys on your
keyboard and how nice it is to sit in your new chair?

All the objects around you that you can perceive and that have been influenced by the human race are manifestations of the smallest, but most sought after, building blocks in contemporary society – ideas. Since the dawn of time and into the new millennium, our fellow human beings have been forming mental images individually or in groups that they have then transformed into a collective reality. In only the past hundred years, our world has evolved from one in which the greater the weight and physicality of objects, the greater their value – commodities, machines, livestock and possessions. The greater, more substantial and tangible our wealth, the more privileged our status in society. From this industrial, economic worldview we are now moving towards a new paradigm in which the intellectual, unique, ingenious and creative is considered more valuable – knowledge, patents, networks and ideas. Ideas and concepts have become such a saleable commodity that there are 'numerous' websites for idea brokers, in other words organizations that trade in ideas.

Researchers, management gurus and political leaders have all begun advocating the ability to innovate as the social factor not only most fundamental in ensuring future growth and welfare but also most vital in creating opportunities for human development and individual self-realization.

The Japanese research institute, Nomura, asserts that our socioeconomic history can be divided into four periods – the Agricultural Age, the Industrial Age, the Information Age and the age that we are now entering, the Age of Creativity. Another proponent of this theory is the former Harvard professor and founder of The Idea Factory, John Kao, who in his book *Jamming* uses musical improvisation as a metaphor to explain the process of human creativity and the trends that suggest we are evolving towards such a creative age. What is actually taking place? And what are the indications that confirm this paradigm shift?

SIX TRENDS THAT ARE CHANGING THE WAY WE WORK

The following six trends in the Age of Creativity, which future organizations will have to contend with and adapt to, have radically altered our approach to work:

1. Ideas are the most valuable commodity in the current marketplace

What are the cornerstones of modern Western prosperity? A great deal of blood, sweat and tears, as well as social organization of course, but at the end of the day it is human creativity that is constantly developing our environment. This has never been truer than it is today. In the current marketplace, a revolutionary idea – whether it be a scientific discovery, an advertising slogan or an IT concept – is worth just as much as endless years of production work. An idea that may have seemed simple and banal at first glance can catapult its originator to financial independence much more quickly than ever before. Ideas are now a commodity, and rights issues for ownership of intellectual capital – in the form of copyrights, patents and trademarks – are the first items on the agenda for international trade organizations. The same way you can buy stuff on eBay, you can today buy different forms of intellectual capital on independent auctions sites like www.ipauctions.com and www.freepatentauction.com.

2. The next phase of the Information Age – creative enrichment of knowledge

In the Information Age, we became experts at creating, selecting and distributing knowledge and information using a wide range of media channels. The IT revolution increased, and is still increasing, our productivity. Now that the Creative Age has arrived, it is time to take this information on board and utilize it – which innovation visionary Debra M. Amidon has labeled second-generation *knowledge management*. Pioneer Professor Leif Edvinsson has been a key contributor in the development of valuation models for intellectual capital, and these are increasing in significance as measuring instruments and steering mechanisms both in the corporate world and in comparing the potentialities of different nation states. In the 1990s, he began developing a system that could value intangible assets, which were often overlooked in corporate financial and annual reporting, but which were frequently the most valuable assets a company possessed – namely the people in the organization and their collective competencies in the form of knowledge, experience, ideas and so on.

3. "Outinnovate" instead of outcompete
Modern-day companies must constantly reinvent themselves to grow and survive. It's a well-known fact that the Finnish company Nokia was a major manufacturer of rubber boots and car tires until it envisioned its future in mobile telephony. And from being the market leader in that category, Nokia now has big problems because of players coming from an industry other than telecommunications, namely the internet industry. Maybe Nokia will have to re-invent itself once again. The corporations of today are waging a perpetual war to become the biggest, the best and the smartest. And on a global level, nations are competing to attract the brightest, most innovative minds to their shores to help ensure economic growth. For example, the Danish government has established a program to transform Denmark into the world's most creative country within the next 10 years. Sweden, for its part, was marketed internationally by former minister of trade Leif Pagrotsky, as "Cool Sweden" to attract creative people. And after the American professor Richard Florida's ideas about the creative class as the main engine of economic growth, there is a constant beauty contest between cities and nations about who is best able to attract the creatives and the bohemians.

It is increasingly becoming a case of possessing the greatest capacity for innovation. In practice, this involves being first to launch new products on the market, ensuring a fast and flexible organization that can anticipate market trends, and developing the smartest, most original marketing measures. It is not until we develop a particularly ingenious idea that we can hit the serious jackpots. And within existing industries there is often a tough competitive climate in which factors such as price, quality and branding squeeze profit margins to such an extent that it is difficult to generate sufficient returns. The ability to innovate has become the single most important development factor and has taken competition in the marketplace to new levels.

4. Design as a competitive factor

For most people, the notion of "working creatively" is synonymous with design. Nowadays, the term "design" is applied in a much broader sense than to describe the work of art directors and industrial designers, which hasn't stopped this particular aspect of creativity from becoming increasingly fundamental. Many people are currently investing in artifacts of all kinds, but basing their investment decision on the design of the artifact and the image the design carries with it rather than the functional aspects of the artifact.

The importance of design is obvious in many areas of business. In the computer industry, Apple has managed to set standards of its own – with accompanying higher prices and profit margins – largely because of its superior product design. Neither the iPod nor the iPhone was technically superior to any other MP3 player or smartphone. The tablet touch screen technology had been used for years before the iPad. The products were not first to market technology-wise – it is simply the design that is the crucial difference from the competitors' products. IKEA delivers the latest trends in furniture design to the mass market. Various consulting firms – such as American firm IDEO and English What If! – show up like stars in the sky and sell design in a box. Fashion company H&M first launches "Cheap Chic" to the people, and then adds a haute couture layer with design stars like Stella McCartney, Karl Lagerfeld and Donatella Versace. And people go crazy.

5. The new generation demands creativity and self-realization

The contemporary workforce has completely different values to those of previous generations. Surveys of recent high school graduates indicate that this new generation prioritizes creative job opportunities over a high salary or job security. In *The Rise of the Creative Class* (2002), Richard Florida writes about the growing mass of creators striving for individual fulfillment, freedom and an entrepreneurial role. In Western societies, this new creative class will soon be larger than the traditional industrial workforce. For creators, a job is not just a means of financial support but also a means for personal development and creative challenge. And according to Florida, the Scandinavian countries and the USA are among the world's most attractive and creative areas as regards the three Ts – talent, technology and tolerance.

Companies need to adapt to this new situation and harvest its fruits, otherwise this new, more mobile, less loyal workforce will move on to greener pastures. And when reading job ads in the daily newspapers, it becomes obvious that organizations are promoting their ability to apply and develop creative talent to attract the brightest minds.

6. Leadership has evolved from controlling to nurturing

Futurist Niklas Lindblad writes in his book, *Window to the Future* (2000), that "the leadership skills of the future, the innovative, will resemble those of a film director's. The collective ability to create is the key to increasing competitive advantages. Managing knowledge and creativity is challenging." And he is most certainly right. The leaders of the future must feel able to relinquish much of the controlling, instructive role that characterized the past, and instead allow colleagues more space, encouraging them to create new ideas and realize their own initiatives. This is partly to enhance the company's growth potential and partly to encourage a better, more satisfied and, above all, more loyal workforce. The development manager of today has the somewhat challenging task of steering a wild tribe of creators in a common direction.

To sum up, it appears that we are entering a new epoch – the Age of Creativity. The philosopher Descartes laid the rational foundation for the Age of Enlightenment with his words: Cogito, ergo sum – I think; therefore I am. The motto for this new age empowered by creativity rather than strict logic could read: Idea habeo, ergo sum – I have an idea, therefore I am! The future lies open to anyone who can understand and exploit the power of ideas and creativity, an energy that can lead both to personal self-realization and financial wellbeing. How have you yourself been affected by the increasing impact and potential of creativity? How does your organization make use of the opportunities and issues thrown up by the forces of innovation? The choice is up to each individual and to every organization – will you harness this powerful energy and lead its development, or will you just get in line and hope for the best?... If you don't make your choice, someone else will take over the wheel and make it for you!

chapter 2

IDEAS, CREATIVITY AND INNOVATION
– or what are they really talking about?

"Whatever made you successful in the past,
WON'T in the future."

Lew Platt, former CEO Hewlett Packard

The world has never been so full of ideas as it is now – they are buzzing like bees in the media babble, creating such storms of stimuli that as individuals we are incapable of processing even a small number. Nor has there ever been such a hunger for ideas as there is today. This has led to a frontier mentality far removed from the days of prospectors wading knee deep in a stream panning for gold. The modern-day prospector is searching for the opportunity to sell an idea to a risk capitalist for a small fortune. This demand for ideas also means that you can open a newspaper at almost any section (job ads, financial pages or social debate) and find three words appearing time and time again – idea, creativity and innovation. Organizations are exploiting these buzzwords to the hilt, but when confronted to describe the practical initiatives that lie behind them, they are frequently caught 'on the hop'. And for organizations that are not taking this development seriously, stagnation lies ahead. On the other hand, for those that learn to understand and apply these terms, the future is most certainly bright.

As we mentioned before, *The Idea Agent* is dedicated to people who would like to become full-fledged Idea Agents. But before you can take on such a role, it would be wise to reflect on these buzzwords that are appearing in more and broader contexts. We should take a moment to consider their meaning and interrelationship.

WHAT IS AN IDEA?

On looking up the word "idea" in a dictionary, one quickly becomes aware of how fundamental it is in our society. It has an assortment of meanings: purpose, opinion, thought, perception, concept and plan, to name but a few. The word has been central to Western thinking since the days of Plato and his expounding the perfect World of Ideas.

In the modern age, it has become something of our mental reality's equivalent to the atom of the physical world – namely the smallest brick in a magnificent construction. At the same time, an idea is worth nothing – it's not something you can put a price on until it's been tested in the real world. Ideas are also the foundation of our future welfare, but impossible to value in their unrefined state and not always receiving their due attention as a result.

WHAT IS CREATIVITY?

Since ideas appear to develop out of creative processes, it ought to be relevant for an Idea Agent to understand the notion of creativity. But defining the word "creativity" is like trying to hold on to a wet bar of soap in the bath. The most convincing and comprehensive research overview on the subject, James C. Kaufmann and Robert J. Sternberg's *A Handbook of Creativity* (2010), comes to exactly the same conclusion. Research into creativity has approached the subject from every conceivable angle (psychological, organizational, biological, social-scientific) and although researchers are convinced that creativity could well be the foremost ability in humans, very few can present firm scientific evidence to support their conviction.

Few words have such an individual meaning and popular use as creativity, and few can cause more ambivalent feelings – joy at one moment and fear in the next. The word originates from the Greek word cre, which means to conceive or to bring into existence, and it is this wider meaning that does it most justice. As an Idea Agent, it is important to accept the fact that creation and creativity in individuals is multi-faceted and difficult to define. Dictionaries frequently define creativity in two ways – partly as an ability (something that can be learned) and partly as a quality (something that is inherited). The qualitative definition only emphasizes the popular misconception that people are either creative or they're not, in other words that we are either born to create or that we were born unlucky and lack the creative ability. Nothing could be further from the truth! All human beings are born with the ability to be creative but not everyone can develop their creativity or has the opportunity to apply their creativity and become skilled at creating.

A huge amount of research has been conducted into the structure of the human brain. One theory – which has more or less been accepted as fact – is Nobel Prize winning American Roger Sperry's reflection from the 1970s that the brain is divided into two functional hemispheres, the left brain and the right brain. According to his simplified thesis, the left half of the brain is oriented to logic, analysis and detail whereas the right half processes the creative, intuitive, visual and holistic impulses. Nowadays, placing too much faith in this theory would be to fall into a very obvious trap, but its most interesting aspect is that it confirms in black and white that creativity is a genuine human commodity.

WHAT IS INNOVATION?

So how are vague terms such as idea and creativity related to the possibly even more diffuse notion of innovation, which has become the number one buzzword of the modern age? Somewhat basically, we could say that a creative process leads to the creation of ideas. But what then is innovation? In her book *The Seeds of Innovation* (2002), Elaine Dundon provides this very accurate, if somewhat strict, business-oriented definition of innovation: "Innovation is the profitable implementation of strategic creativity". This is clearly what companies and organizations are striving to accomplish – an organized, strategic and creative process whose end will be realized and which will achieve measurable results. When one disassembles the definition, one discovers that innovation is:

> **CREATIVITY** – conceiving a new idea.
>
> **STRATEGY** – analyzing the idea's originality and usability.
>
> **IMPLEMENTATION** – setting the unique, usable idea into motion and testing it in the real world.
>
> **PROFITABILITY** – maximizing the unique, usable idea's added value.

To understand this carelessly used term, it can also be useful to emphasize what innovation is not. According to Dundon, innovation is:

Not only new technology

A new organizational structure, a new business channel or a new application for something old are also innovations.

Not business-specific

Innovation is not only needed in major industrial concerns and up-and-comers in the IT and biotech worlds; all organizations should integrate innovation in one form or another.

Not only for R&D departments

R&D departments do a good job for the most part, but an organization that takes innovation seriously applies it throughout the entire organization, from the shop floor to the finance department.

Not only a creative playroom

Creativity involves developing mental and physical environments that are more than just for playing with toys. It's important to have rooms that are designed with the creative process in mind and they should be designed with functional and emotional perspectives in mind.

Not a one-time stunt

A one-day event with speeches from the management will not advance you into the next decade!

Not only creativity training

Coaching professional creative skills is very important, but remember to make training part of an innovation strategy so that these skills take root and are reproduced in day-to-day work situations.

Not only brand-new products

Creating revolutionary new products is a key element of innovation management. But working for continuous improvements in existing products is equally important – possibly even more important since it is through continuous improvements that the majority of companies make their money.

Innovation is clearly not just something that has to be linked to the business community. In recent years, the term social innovation has become increasingly discussed. After Mohammed Yunus received the Nobel Peace Prize in 2006 based on his micro credit initiatives, innovation in the public sector has become the subject of both research and practice.

10 TYPES OF INNOVATION

In any type of business, it's always important to consider at which points during work processes creative phases occur and how these phases should be structured. Innovation can take different forms in different parts of the organization. Each part requires serious consideration and processes should be managed with starting points within each specific situation. Check the following list and ask yourself how your organization manages innovation with regard to each respective area.

A. Product

1. Product features – how good is the quality and functionality of your product? Apple's focus on usability and functionality in the iPhone and the iPad changed the telecom world.

2. Product systems – how can your product be related conceptually with other internal and external products and services? Google's search algorithm has developed into a flora of software solutions that creates value for each other (Android OS, Google +, Google Maps and YouTube).

3. Product services – what services are your customers being offered in relation to the product and how? Singapore Airlines wins lots of prizes in the airline industry, not because they have the best planes, but because they have the best service.

B. Process

4. Core processes – how are internal value-creation processes established on a mechanical and resource-oriented level? Wal-Mart's distribution system from packaging to logistics is extremely advanced and creates an enormous competitive edge.

5. External processes – how are external production frameworks and logistics processes administrated? Most of the big consulting firms like IBM and McKinsey have outsourced their production regarding everything from writing reports to making software to India.

C. Marketing

6. Business models – how does your organization make money? Spotify's business model, both against customers and suppliers (labels and artists), has revolutionized the music business.

7. Brand names – how does your organization manage communication? IKEA's use of their Swedish nationality and yellow and blue color scheme, have created a brand that goes beyond furniture.

8. Sales channels – how does your product reach the customers? The internet has creatively disrupted many industries' sales channels. What is next – the nursing sector?

D. Organization

9. Networks – how is your organization and its value chain structured? Many start-ups concentrate only on research and sale and use partners for the rest of the business. Skype's technological backbone was built in Estonia.

E. Experience

10. Customer experience – how is value added or increased in the form of product experience? What would Harley Davidson be without its motor sound? The next experience trend is trying to patent taste and smell.

Unfortunately, the reality is often such that the only organizations that promote innovation management are research intensive. But the truth is that continuous development is taking place at all levels within all types of businesses, and resources should therefore be allocated for the purpose and encouragement of idea management.

The three notions – idea, creativity and innovation – are essential to the survival of any organization and can frequently be developed along similar lines, regardless of business type. The initial idea phase is the same regardless of whether your company produces interactive computer games or designs sales campaigns for commercial radio. Of course, specialization becomes all the more profound the further into the process one reaches, but for the most part the initial step is universal. And clarifying the various steps often reassures your creative team. Individuals who love brainstorming ideas in an unstructured way will feel that there is a niche for them, and those who prefer a more analytical and tangible approach are reassured that their turn will also come. A simplified model might look like this:

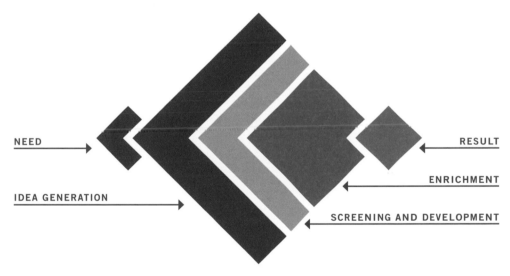

■ NEED

In the need phase, both the project basis and the framework for the end result are established. Carefully executed issue and criteria management will bring success. Moving out of the first phase and into idea generation involves an evolution from analysis to the creative buzz.

■ IDEA GENERATION

In the idea generation phase, various methods and techniques are applied to enable the generation of ideas from as wide a range of perspectives as possible. An incubation period that allows an issue to be digested and processed is valuable, as is the progression from individual to team idea generation.

■ SCREENING AND DEVELOPMENT

When the pool of ideas has been filled, it's time to develop an overview and an understanding of the ideas that can form the basis for screening. The idea development process becomes increasingly discussion based and the number of ideas is reduced but their quality is improved.

■ ENRICHMENT

When only a few ideas are left, the moment has come to develop them from seed to flower, from idea to concept. Enrichment allows idea originators to develop their ideas more deeply, taking them to the next level in the form of graphic descriptions, customer requirements and graphic illustrations.

■ RESULT

The result is a number of concepts that have been developed and evaluated. The winning concept should answer these questions: should we allocate resources to implementing this concept or is the time not yet ripe? Is this a concept that excites us or are its inherent ideas mediocre?

One conventional wisdom is that chaos and creativity are one and the same, but this is a misconception. Generating results through applied creativity that leads to innovation requires a structured process from the need phase to the delivery of a finished concept. This process obviously requires energy, humor and a touch of chaos as well, but it is also structured in a logical and result-oriented manner. Initially, it has the attitude of the curious inventor and the freethinking rebel, only to orientate gradually in the direction of the project owner's pragmatic approach. Our model might appear to be describing idea management and research as a linear process, but anyone who has been involved in the development process knows that it's more often iterative. Issues and solutions that develop during the course of the process can cause loops that will require starting from scratch and finding a new direction. The work of development can seem like a machine whose cogs and wheels are spinning side by side, occasionally disengaging to change places and directions – but in perpetual motion.

IDEA PROCESSES IN DIFFERENT BUSINESS AREAS

Where then can the creative phases of the idea processes in different types of business and organization be found? And have particular businesses contrasting idea processes that require differentiated designs? We should explore a few defined operational sequences in various business sectors and the common challenges within each respective sector.

In professional services such as IT companies and advertising agencies, the skills and reputation of entire organizations are frequently founded upon their ability to create. Often the creative phases consist partly of the establishment of communication platforms with customers, and partly of the internal creative work of the final campaign. The creative process is therefore relatively short. A slogan can usually be generated quite quickly while the creative basis for a campaign can take just a few weeks to develop.

PROFESSIONAL SERVICES

PROJECT SPECIFICATION

COMMUNICATIVE PLATFORM/ WORKSHOP

STRATEGY

CREATIVE BRIEF

STORYBOARD

A common problem in this sector is that the cult of the art director or similar idea engine becomes so entrenched that all creative energies are confined to one individual. As a result, structures can be unclear and difficult to understand for colleagues or new staff, which can easily lead to them being excluded from the creative process.

However, in major manufacturing industries the creative processes can last a great deal longer. In the product development phase, there may be several repeated creative phases, and developing an idea from a Post-it to a finished product on the market may take up to 10 years. In this case, creativity management can be applied in the initial idea process but also in identifying new markets, or in designing the optimal sales campaign. Manufacturing industries frequently fall into the trap of excessively narrow outlooks, or the engineering culture is too inflexible in the idea generation phase. Thus a strict and practical approach to creativity becomes a more important parameter than allowing the creative juices to flow.

In research contexts much of the creative work involves identifying the central issue, or succeeding in highlighting an existing area or issue from new angles. To become a new Einstein or to win a Nobel Prize, one must succeed in finding alternative paths in the drafting phase and if possible new methods of seeking "the truth." An additional critical creative element in the world of research is to find the right means of enabling various financial solutions. Research institutions are often very accomplished at classification and ranking in the selection phase, but the stumbling block is progressing from the analytical "Why?" to the idea-generating "How?"

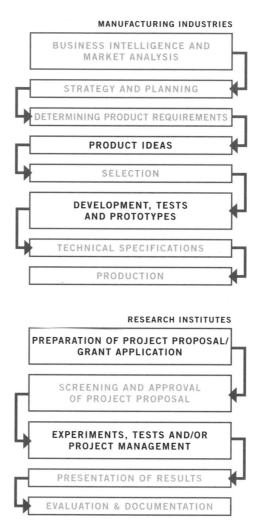

MANUFACTURING INDUSTRIES

BUSINESS INTELLIGENCE AND MARKET ANALYSIS

STRATEGY AND PLANNING

DETERMINING PRODUCT REQUIREMENTS

PRODUCT IDEAS

SELECTION

DEVELOPMENT, TESTS AND PROTOTYPES

TECHNICAL SPECIFICATIONS

PRODUCTION

RESEARCH INSTITUTES

PREPARATION OF PROJECT PROPOSAL/ GRANT APPLICATION

SCREENING AND APPROVAL OF PROJECT PROPOSAL

EXPERIMENTS, TESTS AND/OR PROJECT MANAGEMENT

PRESENTATION OF RESULTS

EVALUATION & DOCUMENTATION

chapter 3

MANAGING CREATIVE PROCESSES

– *or how do I become an idea agent?*

One of the most common mistakes in the realization of idea processes is to allow a creative team free rein in controlling the process as well as the idea generation, instead of delegating responsibility for managing the process to one person. Because assigning a single individual to this function – with total responsibility for focusing on the essential factors of process realization and navigating the path to the end result – is the key to success. We have chosen the title *Idea Agent* for this role. An Idea Agent should be able to understand, harness and supervise the functions of an idea development team, and be able to design processes based on this knowledge. Using a broad range of skills and techniques, an Idea Agent should have the ability to extract the best from each individual in the team and inspire them to creating an optimal end product.

ACTIVE LEADERSHIP IN CREATIVE PROCESSES

Managing creative processes differs from other types of process management to a certain extent, but the key competencies are the same. Supervising processes or running a workshop are grounded on the principle of *servant leadership*. An Idea Agent must assume the position of leader, determining team goals and processes, as well as navigating the team along a well-defined path. In other words, your principal task is not to participate in the process but to manage and facilitate it. An Idea Agent is like a midwife helping the creative team to give birth to brilliant thoughts. An Idea Agent is not a creator of results but the individual that creates the optimal conditions to achieve those results, which means that you must learn to have faith in your team and deal with any uncertainties during the process. This is a skill that doesn't always correspond with the ability to generate ideas. The ideal facilitator does not need to be – and seldom is – the most creatively gifted individual in the organization. Instead, he or she is the person with well developed "helicopter vision" and the instinct for how best to lead a creative team from need to result. The Idea Agent's role can be summarized in these three headings:

THE THREE ROLES OF THE IDEA AGENT

1. GENERALIST AND SPECIALIST The Idea Agent has all-round knowledge of issues and potential end results, but specific knowledge of methods and group processes.

2. COORDINATOR The Idea Agent is the channel between the project owner and the creative team, and the person who organizes resources and expertise.

3. IMPARTIAL OBSERVER The Idea Agent takes an impartial, objective view of different end products, but is continuously focused on navigating the process along the right path to the optimal output.

THE CREATIVE PROCESS

Broadly speaking, a creative process comprises four elements that must be emphasized to achieve optimal results. These are: **knowledge**, **mental environment**, **physical environment** and **method/technique**. A perfect blend of these four ingredients will increase the chances of achieving what is frequently known as "flow," and which in sporting contexts is known as "being in the zone." In other words, the state that most people experience when creativity is flowing freely – time seems to stand still and goals are perceived as an amazing future, attainable and within grasp.

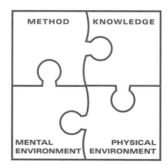

You should begin by considering what methods to use. Whenever the subject of creative techniques comes up, brainstorming is the first example that springs to mind for most people. Brainstorming is a method that was first described in 1953 in the book *Applied Imagination* by Alex Osborn, published by the American advertising agency Batten, Barton, Durstine & Osborn. The method is probably the most well-known group process in the world today. Its essence lies in allowing the mind to explore freely around a central issue, with a team of individuals sitting round a table spouting ideas and a process manager noting them on a whiteboard. When brainstorming was first "launched" as a creativity technique, it represented something of a breakthrough since it focused on creativity as a natural element in the work process. No longer was creativity associated with artists and geniuses only, and the ability to create became an element in the work of the average person. Unfortunately, brainstorming is still the most popular method of idea generation despite the fact that research and practical usage have evolved very much since its conception.

The creativity methods presented in the chapter on idea generation are certainly not rocket science. However, they'll help you redirect your avenues of thought, hopefully outside the conventional boundaries that have evolved as a result of work procedures, social values, peer pressure and the limits of your own knowledge. Actively kick starting thought processes using a strategic approach greatly increases the chances of discovering revolutionary solutions compared to just sitting down and analyzing issues to death using the laws of logic. In the hands of an Idea Agent, the methods themselves become steering mechanisms that can be applied to influence the consciousness of your team. It isn't possible to predict what the final results will be (and therein of course lies the greatest excitement, in fact), but processes can be designed in such a way as to expose and produce quite different perspectives than tackling the focus area using a logical approach only.

The notion that knowledge is vital to creative processes is hardly revolutionary. Whether there's a general issue within your organization that needs solving or a specific product that requires development at an advanced technical level, it is clearly important that a creative team includes individuals with the appropriate skills. It's also an advantage to include people who will remain involved in the project on a long-term basis, those that are most likely to have the specific required competencies as well as the right level of commitment and motivation to find solutions.

But one important aspect that is frequently overlooked is non-knowledge. Finding a brilliant solution often requires an input beyond that of the experts, who are often influenced by the trends of mass media and other herd behavior. In other words, it is crucial to include people in the process who don't sit day in and day out developing business in the particular subject area you need a solution for, for example gifted idea generators, people who can ask those "dumb" questions that the experts have forgotten to ask themselves, even customers with day-to-day experience of the result that the process is intended to produce. In many cases, the additional freedom that "non-knowledge" is usually allowed will advance and liberate "expert" thought patterns at the same time.

To ensure that one has the right people for the creative task, the model below might be useful. The model is partly based on the most famous test on creative personas – the Kirton Adaption-Innovation Inventory (KAI). The theory behind the test, created by the American doctor Michael Kirton, assumes that there are two main styles when it comes to creative work: one is either mainly adaptive or mainly innovative. Adaptors like to develop a smaller number of solutions, while the innovators want quantity. Adaptors require efficiency and enriched ideas while innovators need less concreteness, fewer frames, and fewer rules to get going. Thus, innovators are good to use in the idea generation phase, while the adaptors are best used in the phase of idea development and enrichment. If we let these two parameters be one axis in the model, the other axis will depend upon the degree

The Dolly Method

A Swedish interactive advertising agency discovered this more or less by accident. One day while they were holding a creative session, the cleaner came in. At first, she kept dusting quietly in the background while the team continued their session, which was going very slowly. After a while, she suddenly interrupted them with an idea – there was an awkward hush in the room at first, but her idea was actually very good and it sparked the team into action. Their final proposal was a variation on her original idea and, from that moment on, she was always invited to idea generation sessions. The cleaner's name is Dolly, and since then all the company's innovation processes have been named after her. Try not to forget the motto of this story, nor the curiosity and innocence of Dolly herself!

of knowledge you have about the need. To create a good group for idea generation, one obviously needs the innovation stars who both have the knowledge and are innovative in their thinking. In addition, you need some wise old birds who will be particularly useful in the later parts of the idea process. But do not forget that the idea jokers can be the secret formula for success, namely those who are not necessarily familiar with the subject but who have a lot of great ideas. Where do you find these? Perhaps they are in other parts of the organization than those who

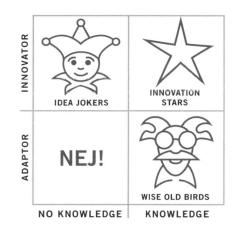

work specifically with this topic. Perhaps they are customers, consumers or even external icons in their fields who are known for their creative force. Those you do not need are the ones who do not know anything about the problem and do not think innovatively – just leave those guys alone.

One important detail regarding the influence of knowledge on processes is the question of how much information team members should be provided about the need itself, both prior to and during the process. Generally speaking, brief outlines are preferable to detailed information. Tasty appetizers should be served up in advance, or a short piece of inspirational background text. The topical introduction at the start of a workshop should last 15–20 minutes at most and include basic facts only. A nightmare scenario is for the presenter to start the day with an hour-long PowerPoint presentation of market research that prefigures the day's prospective results! To begin with, there'll be no creative energy left in the room afterwards, and the prevailing feeling will be that there is no scope left in which to generate new ideas. Participants will instinctively ask themselves: if the presenter already knows the result – what are we doing here in the first place?

Physical environments are a seriously undervalued factor in working creatively. How many new ideas and how much inspiration can possibly be generated in a gray office landscape as compared to a creative playroom? The latter fairly makes your brain cells tingle, doesn't it? Obviously, our message is not that every workplace should look like a playroom, but energy levels will be raised through the use of clear colors and shapes rather than gloomy computer terminals. Observe the mood change in a creative team if they are allowed to stand up, walk around and stick paper on walls as opposed to being forced to sit in chairs chained to their desks. But the design of the room must obviously be managed as well. Environments can easily be adapted to individual issues. When working with product development in the food industry, it might be a worthwhile approach to work in a restaurant kitchen occasionally. If your goal is future housing solutions, a suitable environment could be a high-rise with a panoramic view.

Take plenty of time, absorb and "get into" the room to incorporate its qualities into your process. When designing a practical room for creative processes, it is essential to reflect on these three basic thoughts: work, body and mind.

Work – when setting the stage for an idea process, it's important to adapt your environment to encourage creativity. A good starting point is to empty the room completely and start from scratch, but if this is too impractical, at least do it mentally. Then clear walls, free floor space to enable easy movement and allow as much light as possible to enter the room. Think through how the team will work – in pairs, in groups of five, or all together? – then design desk and chair placements on this basis. Ensure that the whiteboard is clean, that projectors for initial presentations are set up in advance and that all the materials you'll need during the process are prepared and ready for use.

Body – never forget the phrase: "a sound soul in a sound body." Optimizing mental processes requires providing the body with constant necessities such as nutrition, exercise, oxygen and rest. Idea generation is hard, sweaty work. See to it that there is plenty of drink, fruit and candy available on desks for quick energy boosts. Also ensure that team members change places frequently, in the same room or between rooms. Preferably work in a variety of positions (sitting, standing, reclining) so that bodies stay active and don't stiffen up. Physical exercise improves thought processes dramatically. Ensure that the room is regularly aired to maintain a high oxygen level, and you might even consider using particular fragrances to stimulate the mood of the team (lavender calms and citrus refreshes, for example). Ensure that creative sessions don't last too long and don't be afraid to allow individuals to work on their own where they chose.

Mind – key factors in the creative process are stimuli and sources of inspiration. Try to apply colors and shapes on all levels that break conventional patterns and generate energy, high spirits and humor. Humor is something of a magic ingredient in the creative process, each chuckle of laughter giving birth to fresh creative urges and the courage to think in new ways. Humor in physical contexts can be provided using unexpected objects in unpredictable places, which will encourage team members to react and view the world from new perspectives. Why not put vegetables in the fruit bowl, for instance, or leave out round sheets of paper to make notes on? Have fun with it! Also ensure that minds are stimulated with the help of interesting newspapers, books or even curiosities that team members will want to touch or pick up. Anything that stimulates the mind and prepares it for innovation can work. Music is an incredible energy source in a room. The right tunes at the right time can steer a team in precisely the direction you want without the need for instructions. Human moods and mental processes are subconsciously influenced by the magic of music. It's a powerful tool and should therefore be used subtly in the background – it is also essential to interpret your team's character and choose your music accordingly. Experiment at home with your collection and find tunes that you think will have an effect in different situations.

> **What music should you use?**
> Music is underrated as a tool and a stimulus during process management. Try playing some of your favorite artists with the exercises that you prefer. A few tips: Louis Prima gets the blood pumping and suits all ages, Norah Jones works well for team sessions and enrichment, and the piano playing of Eric Satie can induce a reflective mood bordering on meditation in almost anyone. Take a few chances!

Mental environment is probably the factor that fuels the idea process most. By mental environment we mean the creative climate and internal relationships that develop between team members. Most people have experienced idea processes in which the atmosphere was tense and awkward for different reasons, when no one felt especially inclined to contribute anything. Coming up with a seemingly off-the-wall idea in a situation like this is nigh on impossible. Everyone carries his or her own particular inhibitions and mental blocks that need to be released by the Idea Agent for the process to succeed. In other words, energy and humor are crucial to the outcome of idea processes. The groundwork for this is laid using exercises or icebreakers that will enable team members to bond, and which develop a certain initial trust between them. Some even claim that 90 percent of idea development is team building, and though this may be overstating the case somewhat, it still emphasizes the importance of team mental well-being.

When putting your creative team together, ensure that you include individuals with contrasting temperaments who can fulfill diverse functions and roles in the creative process. Of course, conflicts may exist in teams that have worked together previously and these may hamper the work of the Idea Agent, who should also remember that his or her role is not conflict management. To counteract the most common conflicts that can occur, try to ensure that conflicting individuals are not assigned to the same group or next to each other during the idea generation phase. Or simply take the warring factions to one side and tell them that you can see that they're not getting along, but that they should bury the hatchet and discuss their issues some other time because right now there's work to be done.

One of the biggest challenges in everyday professional situations is the capacity to focus and concentrate on the upcoming project. This is why it's important to make sure that team members are genuinely ready for the task that lies ahead of them – both physically and mentally – and that their minds are open and receptive. Generating positive energy and breaking the ice between team members is also crucial to a successful outcome. This is especially relevant if team members have never met before. Try to find out a little about them so that you know what to expect. You should always try to meet the team on their own terms and then gradually lead them over boundaries to innovative thinking. If an Idea Agent can succeed in developing a creative climate within the team right from the outset, the subsequent processes will run much more smoothly.

A skilful Idea Agent will use short exercises and tricks to draw a collective laugh and generate a room full of energy, reassurance and positive feelings. On the other hand, if the team is allowed to remain drowsy and there's no chemistry between team members, the process will develop very slowly and very few ideas will be generated. A skilful Idea Agent can also predict when the time is ripe for an energy boost before exhaustion sets in and the well of inspiration dries up. This ability is grounded on the continuous training of an Idea Agent's ability to read situations as well as preventive planning and simulation. Preferably carry out a simulation of how you think the energy in the team will flow during the process and this will enable you to predict when the peaks and troughs will occur. Then resolve how you can overcome and control these different levels. Contrasting phases will require variations in pace and energy level.

ENERGY LEVEL IN THE TEAM

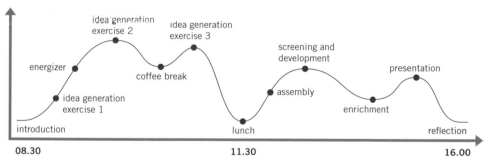

A typical creative team session usually begins with a round of introductions. A skilful Idea Agent is aware that this can be both tedious and unnecessary, and can also cause anxiety in a worse case scenario. Try something new instead of using the same tired old procedures. For example, ask individual team members to describe a hobby that they enjoy, an interesting

Wiping the slate clean

If for one moment you lose focus during idea generation and think about something else, like picking up the kids from day-care, finishing a report before deadline or making a few business calls, a lot of your creative energy will go to waste. Wiping the slate clean is an excellent, simple method of enabling total individual participation.

STEP BY STEP

1. Explain to team members the importance of putting all miscellaneous thoughts and work tasks to one side before the idea process begins.
2. Give each team member a blank sheet of paper and a pen.
3. Ask team members to take five minutes to make a "To do" list of all the things that they have to do when the workshop is finished. Writing down what has to be done will enable team members to put it to the back of their minds.

Tell the team that they should fold up their notes, put them away somewhere and not take them out until the day's work is done. They will now have a reminder of what they need to do later and will be able to focus completely on the job at hand.

place they've visited or even their latest brilliant idea. Encouraging team members to share on a more personal level will enable them to get to know each other more quickly than if they routinely recite what they've studied in school or which rungs they've achieved on their career ladders. If someone would like to know more about another team member, allow their curiosity a free rein but encourage this contact during breaks rather than spending time on it during the actual process.

As well as these simple tips, there are other tried-and-tested methods that can work and that don't feel artificial. Various energy and introduction techniques can be found in the methods section of this handbook. However, it's important as always to think through your exercise choices and the results you're hoping to achieve from them.

Before you select your methods, consider these questions so that you make the right choice:

- **What climate or feelings would I like to develop in the team?**
- **Which exercise best suits my purpose?**
- **What value does the exercise add to the creative context?**
- **How will team members react?**
- **What can I do if the exercise goes wrong and the team reacts unpredictably?**

To support you throughout the planning process, we have compiled a comprehensive lazy 'man's' guide to creative process design that you can keep in mind while you are reading this handbook. The guide can be found in the appendix.

chapter 4

NEED
– or how can we get to the heart of the issue?

AFTER READING THIS CHAPTER YOU'LL BE ABLE TO:
- Identify different types of need and how to differentiate them.
- Define and formulate a focus area.
- Work with tools for need orientations.

All idea generation must have a need as its starting point, and establishing the exact nature of this need is crucial to success. You can generate as many good ideas as you like, but if they solve irrelevant issues then in most cases they'll be of no benefit at all.

Boundaries can be drawn between three types of need in creative processes – film directors need to express an emotion or an opinion, researchers need to identify fresh opportunities in something new or existing, and business developers need to solve their customers' problems more effectively.

In other words, on a general level there are three different types of need, which are listed below in increasing order of abstraction:

1. ISSUE-ORIENTED NEED – an everyday need that arises when, for example, a machine malfunctions, a marketing survey reveals that customers are dissatisfied with a particular aspect of a product, or when a new theme is required for the office party. In other words, there is a defined issue that is tangible and based on a specific need that requires a new solution.

2. OPPORTUNITY-ORIENTED NEED – in this case, there is no immediate issue that requires a solution, instead an opportunity exists to create a future need that people don't yet know that they have. Who could have known that life without cell phones would be impossible in modern society before mobile communication was invented? The opportunity to create a future need can be applied through the utilization of a new material, a new technology, new knowledge on customer behavior, or through research analysis on the healing powers of a particular plant species.

3. EMOTION OR OPINION-ORIENTED NEED – frequently based on the thoughts of one individual, which can be expressed freely in an artistic context and – to all appearances – without limits or rules. Can be of enormous human value and is frequently based on a desire for emotional expression in one particular individual. However, this free artistic need may have the least to gain from the work processes described in this book, even if many idea generation techniques originate out of scriptwriting and research in the gray area between art and more applied development.

Identifying the actual nature of a need is, as we mentioned, one of the cornerstones of all development work. If you start from the wrong point on the map then you will wind up at the wrong destination, having done a great deal of work that was most probably enjoyable and produced inspiring results, but which didn't take the process in the direction the project owner wanted. This is why you should conduct a thorough needs analysis, using the techniques described below for example, to help you form a realistic impression of the actual need. Just be careful not to confuse symptoms with causes.

Your needs inventory may result in you having to carry out a new market analysis or a customer survey. It may also indicate that you should examine your patents on the off chance that you can find the solution there – why invent the wheel if the answer has already been formulated somewhere else? Or you might be required to initiate an idea process to find what you're looking for.

The start of an idea process should be based on the motto "begin with the end in mind." Visualizing the end result and imagining how this can be achieved throughout the process is something that every Idea Agent, in partnership with the project owner, should do prior to process planning and execution. To enable a mental preview of the entire process before start up, the following parameters will need to be discussed:

■ NEED SIGNIFICANCE

How important are the need and the process in comparison with other tasks? Will the project contribute to the development of products that will carry the company into the next century, or will the creative team be generating ideas for a two-week marketing campaign? In the first case, a substantial amount of time should be allocated to the process, and the need can be subdivided into several smaller issues. Hence the process duration will be longer with a large number of workshops of varying character that generate a large number of ideas from different perspectives. On the other hand, the need may be short term in nature and only part of a larger project, in which case only a few days' work will be required. You must decide how important it is actually to achieve a defined goal — in other words, can you afford to throw a few random darts or should you fire a swarm of shotgun pellets to make darn sure that you hit the target of future success?

■ NEED ORIGIN

Where does the need originate from? Who is the real project owner that will develop the end result? Ensure that you know how the need originally developed, or why it was decided to allocate resources to a potential future need. Did the management team issue a directive for the process on the basis of its excellent growth potential in the focus area, does a customer require help with a new advertising campaign, or is there a colleague in the finance department who needs some quick input to help streamline the salary system? Also determine who the stakeholders are and what the need is intended to solve. One risk at this stage is moving too fast and approving a need that seems superficially reasonable only to discover on scraping the surface that it is based on some other underlying issue.

■ OUTCOME CONTENT

It is important to establish what quality standards should be applied for the outcome content. There are often a number of parameters and criteria that steer idea processes, parameters that each idea must be judged by in order for the idea to be subjected to a fair assessment. Should you aim 20 years into the future and try to find revolutionary innovations, or should solutions remain within the framework of continuous improvements and be applicable within three weeks? Should a large number of disconnected ideas be generated or a few tangible concepts? The need will obviously steer the level at which processes are set and ultimately which methods will be used to achieve the end result.

■ PRESENTATION OF OUTCOME

How does the project owner want the end result to look in physical terms? Should the result be presented on 300 Post-its, as three project plans or 10 prototypes made with Lego blocks? There are frequently several questions to be answered before the project owner can form a clear picture of idea construction and the criteria to apply for assessing its quality. Ask project owners what they want in their hand at the end of the process that will inspire them to advance the end result. Examples of end products in paper form can be found in the chapter Enriching and Conceptualizing.

DEFINING THE FOCUS AREA

When you can envision these criteria clearly, it is time to define a focus area that will act as a compass for the entire process. It is important to formulate the focus area as a question whose answer is the result you wish to achieve. This question must be easy to understand and must not include excessive criteria because this will make it hard to apply. The more factors included in the question, the more questions will form in the minds of team members, which will weaken the power of the idea and focus will be lost. If there are several frameworks and factors, it is better to deal with them afterwards.

Nor should a focus area be too broad or too narrow. If the question incorporates too much (How can we save the world?), the result will be too vague. If the question is too specific (How should we adapt the surface structure of product A so that it can be mounted inside product B?), the situation will feel deadlocked and it will be difficult to innovate – the space for ideas will seem restricted. You should also bear the creative team in mind when defining a focus area, as some team members will think that broader definitions are blissfully liberating, whereas others will find narrower definitions more pragmatic and closer to home. Preferably apply future perspectives and descriptive adjectives to strengthen the visionary power of the focus area (How can we convince Generation X to shop in Wal-Mart on a sunny summer's day? What does the ultimate toothbrush look like? Or what will the most popular products within biomedia be in the year 2025?). When you have identified a productive focus area, it is important to consult the project owner and other stakeholders as regards your definition and its accuracy. We can summarize the general process along the following lines:

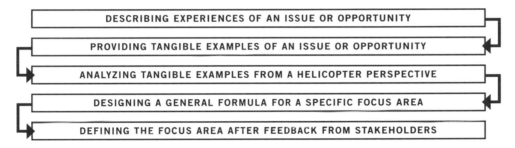

Jumping straight into the idea generation phase without thinking through the direction you want to take will harm the entire process. Even if you succeed in generating a mass of revolutionary ideas, they'll probably be worthless, and the power, desire and faith in the benefits of idea creation will evaporate. Take the process of need definition seriously!

chapter 5

IDEA GENERATION
– *or how can we generate 500 ideas in one morning?*

AFTER READING THIS CHAPTER YOU'LL BE ABLE TO:

- Explain the rules of creative chaos.
- Select the appropriate idea generation methods for your process.
- Implement a range of effective methods for idea generation.

In school we were taught how to think correctly and how to find the right answers. During tests and exams we had to spell correctly and make the right calculations. Nowadays, these tasks are mostly performed by computers, machines and robots. What we still have to solve for ourselves is another type of problem for which there is no text book answer, and for which not one but a million answers exist. This is the realm of idea generation, the geography of development in which atlases – and outer space for that matter – are full of unknown territories for pioneers to explore.

THE RULES OF CREATIVE CHAOS

Idea generation is the essential phase of idea management. It is during this phase that primary idea production takes place and it's here that most of the creative groundwork is laid. There's been much careless talk in many contexts about creative chaos and how complete freedom is the path to success, but even if chaos and freedom are important ingredients, planning an underlying structure is an absolute minimum in a result-oriented process. Like a football quarterback, a brilliant idea usually needs a team and a game plan to succeed. Creativity is often considered a phenomenon for which laws and frameworks don't exist, but the truth is that even in this area a few guidelines will certainly help.

Before starting the idea generation phase, it's worth introducing your creative team to the following principles for the **DOs** and the **DON'Ts** of idea generation.

CREATIVE DOs

■ COME FLY WITH ME

Letting go of inhibitions is crucial in all idea processes. In the end, it's those off-the-wall ideas that will help you develop the new approaches and perspectives that can lead to the novel and fascinating. Encourage team members to develop each other's ideas and apply them as springboards to the next level of team energy and a rewarding result.

■ GENERATE AN IDEA POOL

It must be rather obvious by now that it's quantity rather than quality that is the primary focus during the idea generation phase. At the start of the idea process, it's important to generate as many ideas as possible in as many categories as possible so that you can establish an idea pool for further development. Try to estimate in advance how many ideas you want to produce so that the team has a clear and definite goal that can be achieved and measured up to. Every new idea contains the seeds for hundreds more!

■ WRITE EVERYTHING DOWN/VISUALIZE THE IDEAS

While the stream of ideas is flowing freely, it's easy to forget to document what is being said – to actually preserve the idea capital that's being generated. Make sure that everyone has access to paper, Post-its and pens so that thoughts are written down and not overlooked. If you don't appoint a team member as secretary – which is frequently a misuse of resources anyway – then you should point out that each team member is responsible for their own ideas. It doesn't matter if some ideas are duplicated. And ideas don't need to be recorded in detail during the idea generation phase itself, just a few key phrases will do. The Idea Agent must then ensure that time is allocated after the exercise for team members to revise what they've written so that it can be understood by the others.

Quantity is Quality

The importance of mass idea production cannot be overemphasized. A majority of surveys indicate similar results – only a few ideas per idea generation session will have real impact and the more ideas produced the greater the chances of a revolutionary result. The research results presented below are based on statistics from the software company Imaginatik Inc, who develop idea management programs.

2% – 3% high impact, high-quality concepts
10% – 25% ideas worth developing
20% – 45% ideas that have already been generated or implemented
5% – 10% duplicates
15% – 30% "non-ideas" such as expressions, observations and requests

■ JUST SHOOT!

In the world of psychoanalysis they often talk about "thought control," in other words the mental policeman that sits in our head stopping us from saying something dumb, a form of self-censorship whose laws are founded on many years of learning and experience of social rules and norms. Try to encourage team members to give their mental police time off during the idea process and emphasize the importance of expressing ideas verbally as soon as they spring to mind. Be spontaneous, impulsive and speak before you think for a change!

CREATIVE DON'Ts

■ CRITICIZING AND ASSESSING

To start putting down another person's ideas as early as the idea generation phase is devastating to the creative process. This is the most common of all the diseases in the category "ideophobia". Idea assessment will follow the process right through to realization, but it should never take place during idea generation. Criticism diverts the mind into counterproductive lines of thought and blocks ideas of all kinds from emerging. The Norwegian consultancy firm Stig og Stein has a good rule that will help avoid negative criticism: every time you feel the urge to say "No..." or "But...", you should force yourself to say "Yes, and..." then enrich the idea that you're about to criticize.

■ THE BOSS ALWAYS KNOWS BEST

In many organizations – especially those founded on an authoritarian hierarchy – it is frequently the boss's thoughts and ideas that count. The point of most meetings is to tell the boss what they want to hear. In an idea generation session this is always fatal. Persistent political correctness and trying to please will not benefit the development of any organization in the long term. This is why you should have a word with the boss about his or her place in the team during the idea management process. Should he or she participate or not? If she does take part, it should be on the understanding that it's on an equal footing with the rest of the team and that she might even consider toning down her role to allow the other team members more creative space.

■ TAKING TURNS TO EXPRESS YOURSELF

Democracy is a great institution but it doesn't work especially well in idea generation. Being too polite or putting up your hand and waiting patiently to get a word in will completely stymie that all-important creative spontaneity. When you're spouting ideas, the laws of anarchic freedom must apply. It's OK to interrupt your colleagues.

■ ONLY THE EXPERTS ARE ALLOWED TO CONTRIBUTE

Particularly in knowledge-intensive development processes, it is standard practice always to consult experts for advice and to carry out a thorough analysis that will help to find desired answers. But if you're keen to create truly revolutionary ideas, it's very unlikely that you will succeed if you put 10 like-minded business developers from the same unit on the job. You should obviously listen to the advice of experts and use analysis for input during need presentations, but when the idea generation process is in full swing, everyone should be given the opportunity to speak their minds. Laymen, customers or consumers can all make a worthwhile contribution by re-asking basic questions or thinking outside the self-imposed boundaries of the subject specialist.

■ SCORNING OFF-BEAT IDEAS

How many times have we endured the experience of a really negative person making faces at every idea we come up with? If this type of character is given too much space during idea generation, the entire creative process will grind to halt since the team will be too afraid of making a slip up and being subjected to a look that brands them as a prize jerk. As an Idea Agent, it's very important to keep tendencies like this in check, to encourage wacky ideas and to help negative personalities understand that this type of body language isn't welcome. In the idea generation phase, it's uncool to be boring and hip to be a bit of a nut!

■ ANALYZING AND INTELLECTUALIZING

In creative contexts, no idea needs to be well thought out before it's expressed. Anything is possible at the start of the process. In due course, all ideas will be developed, screened and assessed using every feasible and unfeasible criteria, but at the start of the process there are no limits. In other words, try not to get caught up in your thoughts because you're uncertain that your solution is practical or because there is no established market research that indicates that the idea might work. Dare to be a visionary!

THE PATH TO CORRECT IDEA GENERATION METHODS

When guidelines for creative freedom have been established, the time has come to present the methods that you're intending to use. As we mentioned before, the vast majority of creative groups still use brainstorming as their primary means of creating ideas. But this is just one method among dozens that can contribute to generating ideas from different angles and with varying content. The methods included in this chapter appeal to the logical, the associative and the role player in human creativity. Certain methods will work very effectively for an individual with a pronounced ability for visualization, while another personality may require clear connections from A to B, and a third will revel in being allowed to play the devil's advocate. General classifications of methods are hard to make because different stimuli and thought processes have different impacts on different individuals. Perhaps it's here, in the creative mystique, that the working Idea Agent's fascination for idea processes ought to lie.

Having decided which idea generation methods are to be used, as an Idea Agent it is important to ask yourself these questions to help realize optimal results:

1. HOW MANY IDEAS SHOULD BE PRODUCED AND OF WHAT QUALITY?
The results of the idea generation phase can differ depending on the methods you use and the process focus. Is your goal is to produce 500 ideas of two to three words on Post-its, or 30 prototypes made out of matchsticks? Should these ideas be inside the box (i.e. logical improvements within an existing framework) for which you can apply consistent thoughts as a sound basis for your result? Or is there no remaining space left for thoughts inside the box and it's time to start searching for something novel outside the laws of logic?

Ensure that methods are applied that will encompass different parts of the focus area – both logical solutions inside the box and more exploratory ones outside prevailing models. Generally speaking, an idea pool is more productive because it increases the chances of finding new seeds and encourages interesting cross-breeding.

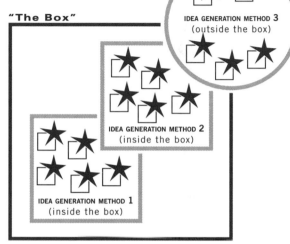

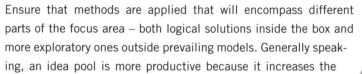

2. HOW CREATIVELY MATURE IS YOUR TEAM?

Before an idea process starts, it's a good idea to form a general impression of your team members. What cultures exist within the team or organization? Are team members outgoing and broad minded, or cautious and fearful of leaving the security of their offices? To connect with an individual and get the most out of them, you must meet with them on familiar territory. If you move too fast and cross the threshold of their personal integrity or space, they might resist you. Plan the process carefully, based on what you think team members can cope with, but don't forget to challenge them to think outside established norms as a certain degree of boundary crossing will stimulate creativity.

The names in the chart refer to the methods described in the Method section.

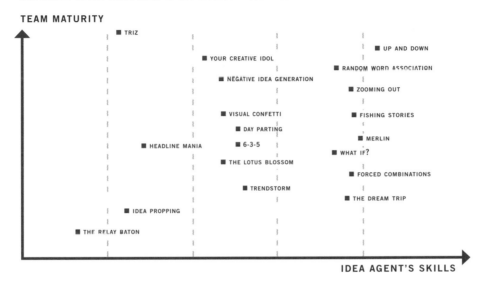

TEAM MATURITY

- TRIZ
- UP AND DOWN
- YOUR CREATIVE IDOL
- RANDOM WORD ASSOCIATION
- NEGATIVE IDEA GENERATION
- ZOOMING OUT
- VISUAL CONFETTI
- FISHING STORIES
- DAY PARTING
- MERLIN
- HEADLINE MANIA
- 6-3-5
- WHAT IF?
- THE LOTUS BLOSSOM
- FORCED COMBINATIONS
- TRENDSTORM
- THE DREAM TRIP
- IDEA PROPPING
- THE RELAY BATON

IDEA AGENT'S SKILLS

3. HOW MANY METHODS SHOULD BE USED AND IN WHAT ORDER?

When you've determined what is to be achieved results-wise and you've assessed the creative maturity of team members, it's time to decide how many methods to apply and in what order they should be applied. As a part of a major development project, you might need to hold three separate workshops, each using three idea generation methods for three different groups. Or a half-hour activity using only one method may be enough to solve a less complex issue. Consider alternating between methods with contrasting structures and stimuli to include as many perspectives and personalities as possible. And remember that working with ideas is tough – no matter what people claim – and that a maximum of three to four techniques in a row, or idea generation for a total duration of three to four hours, is the most your team will have the energy for. The creative part of the brain will become exhausted and needs oxygen, energy and rest. On the other hand, this doesn't mean that other elements of the process can't be dealt with instead.

It only remains now to pick and choose between methods. An important parameter for an Idea Agent during method selection is that you should like, feel comfortable with and have faith in the potential of your methods to achieve worthwhile results. Though you can never entirely predict what will happen during a process, you must feel confident that the methods you are applying can contribute positively to results. The better you become at managing processes the more you'll be able to design your own methods or improvise freely using existing ones.

SO WHERE ELSE CAN YOU FIND IDEAS?

The focus of this book is the process of creating ideas, but be sure not to do unnecessary work. Of course one must not reinvent the wheel, if it is already invented. In today's global world there are often a lot of ideas to seek, receive or buy both inside and outside the organization. Especially in large multinational organizations, it often pays to search in internal idea management systems or consult parts of the organization as to whether they have the solutions you are seeking. In addition, initiatives on open innovation have become increasingly popular. The theory behind open innovation is that you cannot have the entire world's talent within the organization, but must learn to listen and cooperate with the world around you in order to create the best ideas. One can, for example, work with the customers and ask them to help with ideas, something that especially the internet industry has been good at. The games company Massive Entertainment has gained great value from their gaming customers. Massive Entertainment has opened up some of its source code to the gamers which has resulted in them creating their own maps and worlds which they share with others in the gaming community around the game *World in Conflict*. Massive Entertainment pays nothing for this but it creates an enormous value for the company. But you can also buy open innovation solutions from companies such as NineSigma and Qmarkets, which are both working with networks of researchers and innovators. You send them a description of your need, and after some time you get some answers from experts around the world.

And finally – don't forget at some point to allow team members the opportunity to note down the ideas that they may have brought with them. 'Old favorites' might just work fine!

chapter 6

SCREENING AND DEVELOPMENT

-or how do we clear out the weeds?

AFTER READING THIS CHAPTER YOU'LL BE ABLE TO:
- Argue in favor of different approaches in the idea development process.
- Understand how to establish screening criteria.
- Work with different screening methods.

By now there should be hundreds of ideas stuck to the office wall in the shape of Post-its, loose sheets of paper or sketches. The huge number of ideas and the intellectual capital that the team has succeeded in generating are overwhelming, and you have to catch your breath when you realize how much hard work lies ahead in organizing and developing the material. It must seem like you're trying to clear the weeds from an enormous field of carrots, but you just have to get down on your knees, roll your sleeves up and start clearing around the first carrot you see. Because spraying the field would kill the whole crop – all the carrots have to be weeded individually and with the same loving objectivity.

But to get back to the idea pool, an unfortunate but typical post-idea-generation scenario is for one individual to stay behind on his own, figuring out what's been written on the various scraps of paper and assessing which ideas are the best. In situations like this, it probably feels as if you are suffocating in an avalanche of ideas. Tiredness sets in just as soon as you try to decipher the first unrelated ideas in the pool and instead of making a thorough inventory, you choose a few of the notes you've written yourself because at least you can read them and understand what they say (plus of course they're your own favorites). The disadvantage with this approach is that a huge quantity of intellectual capital will go to waste, forgotten in a drawer at best or discarded in the trash can at worst.

To avoid this scenario, you must make sure that the entire team continues working on the process in the phases after idea generation. The tasks that follow idea generation are not as spontaneous or creatively inspiring as idea spouting, and require a more introspective and analytical approach. Nevertheless, they may still lead to new ideas being hatched as a result of the interesting mixes of areas and thought-seeds. It is interesting to note that team members who may have seemed a little withdrawn during the idea generation phase may now come into their own. And once again, it's important to emphasize the significance of a heterogeneous team in successful idea processes.

The screening and development phase is a case of putting more flesh on the bones of ideas and viewing the idea pool from a helicopter perspective. It's a phase that can be completed in a few hours, but can also last a few days or even months if major research is needed to evaluate the potential impact of your ideas.

You will frequently start off with several hundred ideas – some of which may consist of only one word and seem worthless at first glance, while others may be well defined and even described graphically – that will ultimately be whittled down to a small number of more tangible concepts. Sometimes the team will discuss subjects that are purely visionary in nature: Which path should the organization take in the future? There may be differences of opinion and heated debate. It is crucial that the Idea Agent knows instinctively when to allow these discussions to develop. It may well be an issue that the team really needs to discuss. On the other hand, if time is short, you might need to step in and steer the process onwards so that its pace doesn't slow too much.

In any case, it is essential to make contact with the project owner by one means or another. Either the Idea Agent will have had a discussion with the project owner at an earlier stage in the process and have the selection criteria available, or the project owner can be encouraged to communicate directly with the team. Team members will then have the opportunity to ask questions and receive guidelines for the values or factors that will steer the remainder of the process.

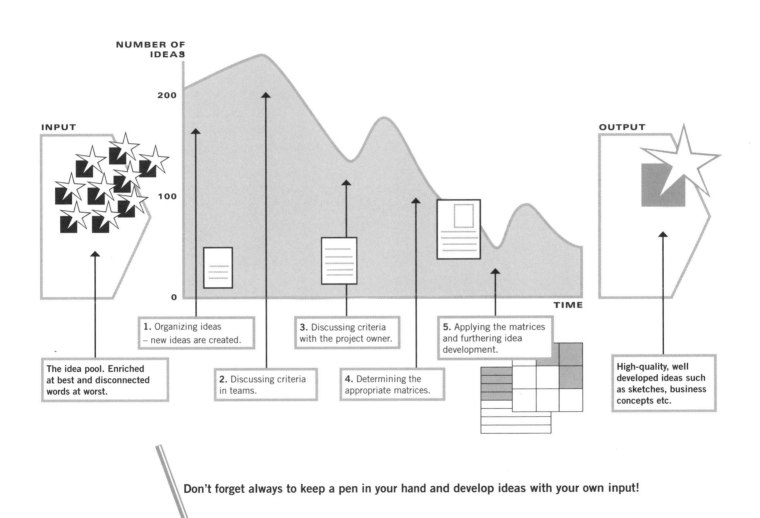

NUMBER OF IDEAS

200

100

0

INPUT

OUTPUT

TIME

The idea pool. Enriched at best and disconnected words at worst.

1. Organizing ideas – new ideas are created.

2. Discussing criteria in teams.

3. Discussing criteria with the project owner.

4. Determining the appropriate matrices.

5. Applying the matrices and furthering idea development.

High-quality, well developed ideas such as sketches, business concepts etc.

Don't forget always to keep a pen in your hand and develop ideas with your own input!

In the organizational phase, you can choose to apply one of two primary work processes that are similar in nature, but which have one basic difference. This difference may appear superficial but it can have a profound effect on the end result. You can either opt to focus on the potentiality of individual ideas, or you can choose to concentrate your ideas into clusters with a collective impact on the end result (by "clusters" we mean a group of ideas within one particular subject area). If you choose to focus on individual ideas, each idea should be assessed individually and on its own merits, regardless of the subject area it concerns. Each idea originator will therefore have more opportunity to develop their own "babies," and the project owner will have less influence in the initial stages. This approach can be beneficial if the focus area is relatively broad and the project owner has relatively few predefined require-ments for the end result. On the other hand, if the project owner has predefined a segment for the focus area that the result should address, a cluster method may be more appropriate. Each idea then becomes a part of a cluster or a subject area, with each area being assessed on the basis of its relevance to the issue. In other words, no initial attention is paid to individual ideas, instead the ideas contained in the clusters considered most relevant after evaluation are then developed further. Within these two approaches, you can choose to apply various combinations of screening and development methods to find the optimal process. Here is a step-by-step description of the two approaches.

THE CLUSTER METHOD

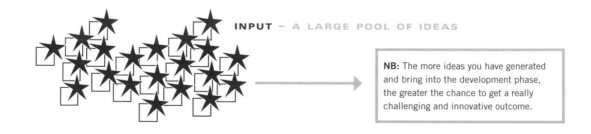

INPUT – A LARGE POOL OF IDEAS

NB: The more ideas you have generated and bring into the development phase, the greater the chance to get a really challenging and innovative outcome.

Step A1 – create clusters

Reflect on your ideas and decide which belong in the same subject area (for example, content, form, campaign, external communication). Assign different titles to your clusters. Make sure that you read what's written on the Post-its carefully, not too quickly or casually. Important information may be lost if the ideas are not interpreted clearly. Try and work in pairs to ensure that at least two pairs of eyes are involved in the process. Put the ideas up on one wall and transfer ideas one by one to another wall with headings where clusters can be formed.

When all the ideas have been organized into clusters, take one step backwards and ensure that cluster distribution is practical and tangible. If certain clusters are too large, it would be wise to reorganize them. Avoid vague cluster titles such as "miscellaneous" and "indefinable." Ask your team members if they're satisfied with the cluster titles or whether they should be redefined for the sake of clarity. Also ask yourself whether there are enough ideas in each cluster. If more ideas are required for a certain cluster, schedule more time to generating more ideas within those clusters.

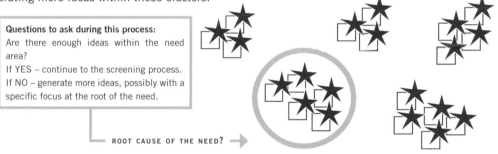

Questions to ask during this process:
Are there enough ideas within the need area?
If YES – continue to the screening process.
If NO – generate more ideas, possibly with a specific focus at the root of the need.

ROOT CAUSE OF THE NEED?

Step A2 – evaluate the importance of clusters in relation to the focus area

Now discuss how to weight the importance of respective clusters as solutions to the initial issue. Ensure that those with the greatest potential impact are placed closest to the center of the target area and give them the highest grade. The clusters with the least relevance should be given the lowest grade. For example, grade the clusters from one to 10, where one is the lowest grade and 10 is the highest. Of course, several clusters can be given the highest grade if they are all considered essential to the issue. At this point, discussions may develop as

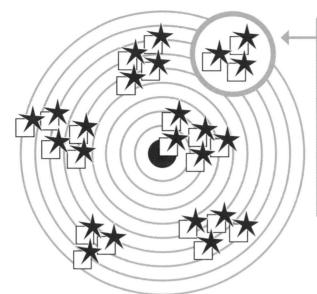

Questions to ask during the process:
A. Can less important clusters be used as inspiration sources for new thinking and fresh ideas in the core area later in the process?
If YES – why not delegate a "guerilla force" to develop the less important clusters, parallel to the main process?
Unexpected brilliant ideas are often generated on the fringes.
If NO – continue according to plan.

B. Are there enough ideas to work with later in the process?
If YES – continue according to plan.
If NO – generate more ideas!

to the importance and grading of different clusters and this is why it's important to be able to contact the project owner.

Step A3 – set limits

When all the clusters have been graded, the time has come to decide where to draw the cut-off line for progression. You may decide that all clusters graded five or above are worth developing, or if time is short only those that have been given the top grade. Make sure that there is some degree of separation and that some of the clusters get discarded – you must be able to let go of a few favorites and move on. Please note that the "discarded" clusters always get another chance! They should be examined for any hidden pearls that can be advanced despite the fact that the cluster as a whole didn't qualify. Pick theses ideas out and add them to the finalists. When the decision has been made, it is time to move on to step 4.

THE IDEA METHOD

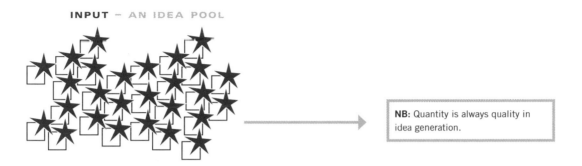

NB: Quantity is always quality in idea generation.

Step B1 – survey the situation

Make sure that team members have written down their ideas clearly and concisely. Put all the ideas up on the wall or on the floor or a table to enable the best possible overview, then mingle for five to 10 minutes so that you can develop a better understanding of the idea pool. Team members can also create clusters to enhance their perspectives but this is not a requirement. i

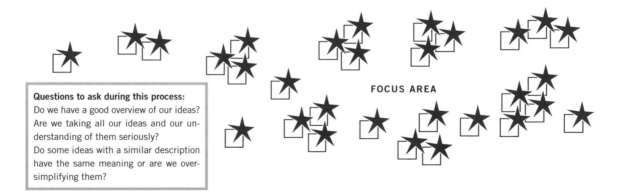

FOCUS AREA

Questions to ask during this process:
Do we have a good overview of our ideas?
Are we taking all our ideas and our understanding of them seriously?
Do some ideas with a similar description have the same meaning or are we oversimplifying them?

Step B2 – decide how many ideas should be developed

Estimate how much development time is available and then decide what proportion of the total pool should be developed accordingly. Should it be half or only 25 percent?

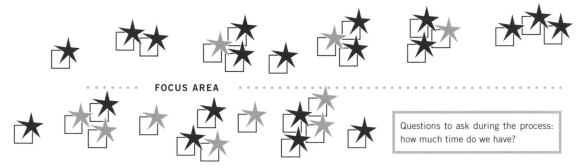

FOCUS AREA

Questions to ask during the process: how much time do we have?

Step B3 – select your favorites

If you decide that half the ideas should advance, each team member should pick half of their favorite ideas and transfer them to another board, desk etc. No consideration should be taken for possible clusters and all good ideas should advance.

YES!   Not right now...

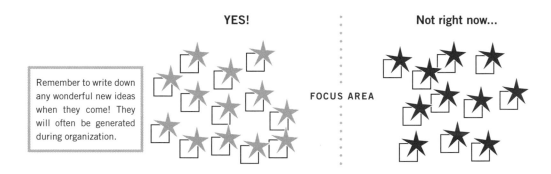

Remember to write down any wonderful new ideas when they come! They will often be generated during organization.

FOCUS AREA

When all the ideas are in place, continue to step 4.

SCREENING

Step 4. – what are your essential selection criteria?

If you haven't yet received the essential selection criteria from the project owner, they are definitely overdue. Ask the project owner to specify and prioritize the five or six most essential criteria for product development. If the project owner is unavailable for some reason, team members should establish and rank feasible criteria on the basis of their own working knowledge. Criteria management is complex and includes a number of tricky interpretations and strategic aspects. However, it is important not to get bogged down in discussions to prevent an energy drain.

Typical criteria are:

- Feasibility
- Innovativeness
- Time to market
- Cost
- Uniqueness

- Customer value
- Risk
- Appeal
- Operating margin
- Market potential

These criteria will steer the process and your idea selection for the matrices or other idea screening methods that you choose. If you can't obtain a specification for these, let the team discuss and determine the selection criteria that they believe to be most important.

Step 5. – which matrices or idea screening methods should I choose and how many?

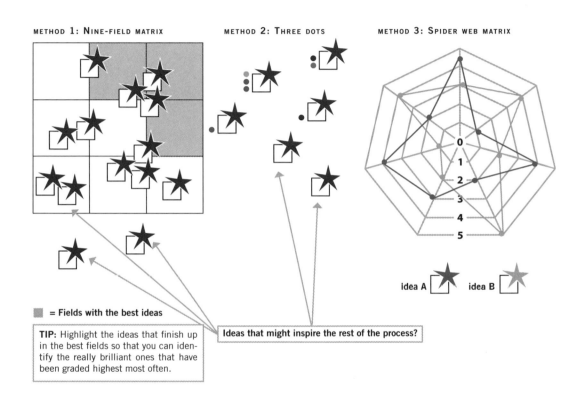

METHOD 1: NINE-FIELD MATRIX   METHOD 2: THREE DOTS   METHOD 3: SPIDER WEB MATRIX

idea A   idea B

■ = Fields with the best ideas

TIP: Highlight the ideas that finish up in the best fields so that you can identify the really brilliant ones that have been graded highest most often.

Ideas that might inspire the rest of the process?

After the initial basic screening, the time has come to apply a range of systematic screening techniques based on your defined criteria, with the aim of sifting the idea pool further. The standard approach is to use a four-field or nine-field matrix, but there are a handful of other options that you can apply if you find the use of matrices too rigid. Be careful not to apply matrices whose axes are too similar to ensure that results are not meaningless or difficult to interpret.

It may well be that several winning ideas have already crystallized after your initial matrix, but if you're addressing a complex issue that requires a thorough approach, you had best use several screening methods to ensure that you make the correct selection. It is also extremely important to try to develop your ideas continually as you insert them into the matrix! If you don't do this, or make new combinations while working with your ideas, you will advance from the screening phase with exactly the same ideas that you started with. Add, add and add relevant information all the time! This is crucial to your chances of success. For example, if an idea is extremely risky financially, think it through and make a few adjustments that will reduce the risk and give it a higher evaluation.

If you're working on a process of lesser significance, you may decide to wind up the screening process and advance with your best ideas to the enrichment phase. But if the project is a major commitment that requires a thorough appraisal, you can continue the screening process using more precise quantification methods and measurements. On the other hand, you may prefer to build your ideas first to increase your understanding of them before you advance to step 6.

Step 6. – quantification methods

If you feel that it's absolutely necessary to continue with the screening process of say the three to five best ideas, there are several quantification methods you can apply. These will provide you with a more precise evaluation in the form of a numerical grade and a total figure. Should you wish to weight the criteria internally and quantify the potential of ideas mathematically, a much-used technique is the Kesselring method. However, Kesselring involves a certain degree of preparation and that needs available time for longer periods of development. To help visualize a graphic image of the strengths and weaknesses of your idea, a spider web diagram can also be very useful.

chapter 7

ENRICHING AND CONCEPTUALIZING

– or how can we dress up our ideas?

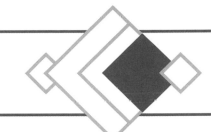

When your basic ideas have been screened and developed, the moment has come to advance them from idea to concept. This will involve asking the really tricky questions, as well as developing a far more in-depth picture of the scattered thoughts that have formed the core of your idea until now. There is often a grand vision in an idea originator's mind that hasn't yet been expressed or documented, which is why it's important to conceptualize your ideas so that the rest of the world can understand them. We have all had our moments of inspiration, but putting this inspiration down on paper in a coherent and presentable form often proves quite a challenge. And if you can't conceptualize your idea so that other people can understand it, this raises the question: how can it be all that inspirational?

Depending on what has been decided with the project owner, there should be a number of ideas remaining in your idea pool ready for enrichment, usually somewhere between three and 10. And as during earlier phases, it is essential to alternate continually between individual and team sessions with the aim of maximizing the potentiality of each personality in the team. One recommendation during this phase is to allow team members to start the enrichment process individually because they'll often have their own favorites that they can't wait to work on. However, after allowing them a little time to themselves, they should be organized into groups so that they can thrash out their ideas and enrich them with each other's thoughts, suggestions and smart questions.

DIFFERENT TYPES OF ENRICHMENT

So what form should concept enrichment take? This clearly depends very much on what type of process you're working with. If your process is in support of an advertising agency, the desired result could well be a slogan on a sheet of paper, a graphic concept, or a project plan for a client campaign. Each of these three results will require different approaches to bringing the idea to life. If you are developing a new machine in a manufacturing company, enrichment may consist of a mechanical sketch and a simple prototype, in addition to issues relating to materials, cost calculations, technical solutions and the impact assessment for stakeholders in the value chain. In a research project, the result could be the development of a new research method or network. In which case, enrichment will probably consist of a goal, potential stakeholders, a prospective project manager and the challenge of obtaining a grant.

When the first step in the enrichment process has been completed, it may become necessary to conduct additional quality evaluations of the concept, and you may also need to take a step backwards and repeat the screening and evaluation process. New focus areas, new information or new ideas might have developed that need addressing, which will sometimes mean starting the entire idea process again from square one. Loops like these are unavoidable and should be seen as the constructive consequences of the continuous development process that we are all a part of.

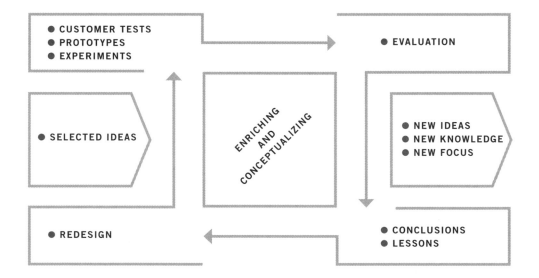

TESTING IDEAS

Making fast and reasonably OK prototypes or pilots has become increasingly important in the development of new ideas. Why? Well, the time window to launch new products in the market is becoming narrower and the cost of developing – and above all launching – new products has increased. Therefore, it is crucial to showcase concepts as soon as possible with as little financial effort as possible. And thus quickly decide upon go or no go. So, how do you quickly create a selling prototype of an idea? Either you hire a visualizer / designer to help you do some sketches. Or you teach yourself 3D-software and use it. Or you buy some hardware for prototyping. A visualizer is a person that can illustrate how a product or service might look, make a sketch on a mood board or perhaps a CAD drawing of the concept in its infancy. A great free software that can be used for 3D-modeling is Google SketchUp. Rapid prototyping is becoming more popular now that 3D printers are getting cheap to buy and use. One can now produce small physical prototypes in plastic for a reasonable price, making the concept much clearer, more graspable, and easy for others to understand. And one should never underestimate the power of being able to show someone a physical product or an advanced graphical model. Something you can touch and feel makes a big difference in relation to a block of text.

To let potential future users test different ideas or concepts at an early stage is also important. Working with focus groups and beta testing are becoming increasingly common in today's development work. However, note what type of question you ask and what kind of innovation you're working with – focus groups work well in the evaluation of ideas for continuous improvement, but worse when testing paradigm shifting solutions. The masses are often not receptive to revolutionary ideas. If you want to test something that is really out of the box, then you have to work with so-called early adopters, persons who are quick to pick up news and set trends. The classical development process was based on thinking that everything should be 100 percent complete and ready before the launch of a product. That is changing. The trend goes toward releasing products and concepts early in the development process, in order to learn from the users, minimize time to market and pick up on market trends quickly. Different approaches to this idea have been discussed in terms of agile design and rapid prototyping, and it has also been launched by the American entrepreneur Eric Ries as the concept of the "lean startup".

FROM ACTIVE PROCESS MANAGER TO PASSIVE HELPER

If you've done your homework and have already established the format for the end result during the initial stages of the idea process, then the enrichment phase is an easy ride for an Idea Agent. You'll be aware of the strategic aspects that enable an understanding of the concept and the likelihood of these options achieving success. And you'll have designed the documents you want to use and maybe even contributed with materials for building and testing prototypes.

The role of the Idea Agent now becomes less crucial to the process, with an increasing amount of responsibility transferred to the team members themselves. The further into the enrichment process you reach, the more specialist competencies will be needed to make a contribution meaningful, and process management becomes more a question of making sure that all the required materials are available, that suitable refreshments are provided and that the schedule is maintained. It is obviously important to make sure that high energy levels are sustained during the process and that the end results are as precise and coherent as possible so that you can pass them on with a clear conscience. Sustaining energy levels may simply mean airing the room by opening a window, but it can also involve calling in a few outside experts to invigorate the team and act as sounding boards. One useful method is to let a few "free operators" with specific competencies mingle and ask questions of team members, which can provide fresh perspectives on the thoughts and concepts that they are trying to generate. These experts can also use their skills to assist in any issues that arise during the process.

To enable an insight into the variety of enrichment solutions available, here are a few examples of simple graphic tools that will enable both verbal and visual concept descriptions. Most of them are fairly generic and can be applied to most categories, but the relevant questions must obviously be formulated to suit your process. You could also mention to team members that these idea templates don't need to be applied to the letter; they are more a means of developing concepts from as many perspectives as possible.

A basic version of a graphic template, which can be applied for idea enrichment in which results do not need a detailed description and for which time is limited.

name of idea:

brief description of idea:

how is the idea innovative?

what customer need does the idea fulfil?

visualize/sketch the idea:

idea originator:

A graphic idea tool that explores various aspects of idea formulation in detail. This template is useful if the project owner requires an extremely clear concept that is well defined on many levels.

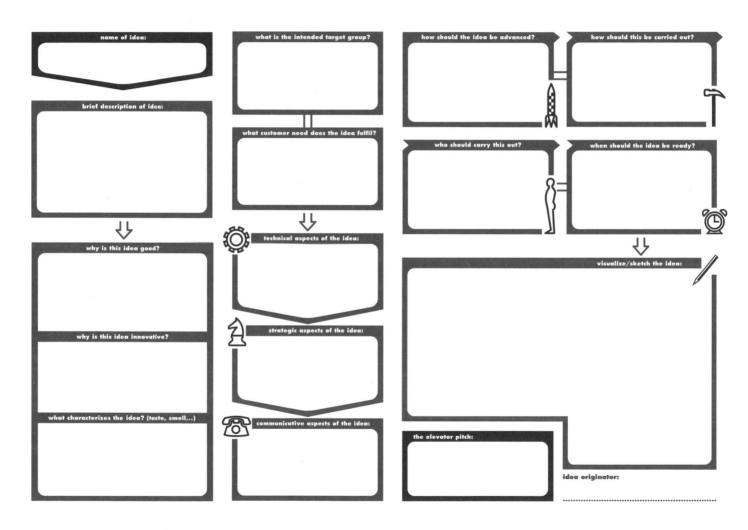

An idea tool that focuses on idea visualization. This template is practical if design sketches or similar are essential to idea assessment.

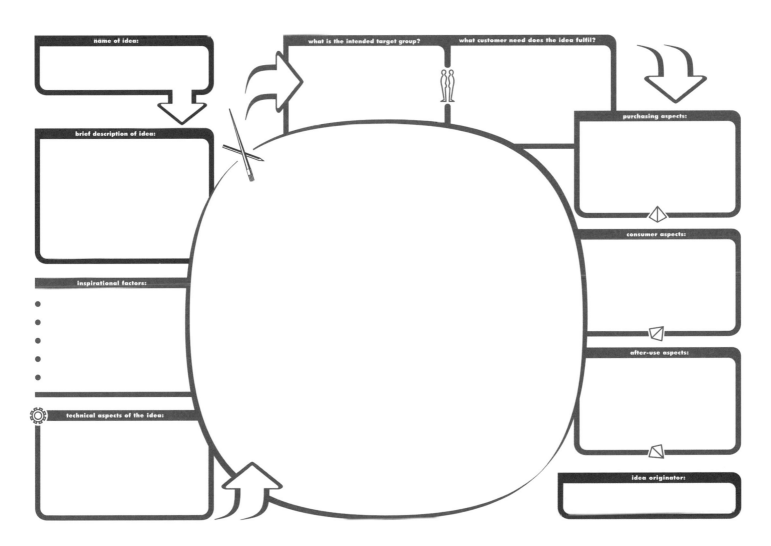

name of idea:

what is the intended target group?

what customer need does the idea fulfil?

brief description of idea:

purchasing aspects:

consumer aspects:

inspirational factors:

after-use aspects:

technical aspects of the idea:

idea originator:

An idea template that applies a traditional SWOT analysis as an element in the enrichment process, as well as an additional assessment matrix. The fact that team members themselves carry out part of the appraisal can help the project owner during further development.

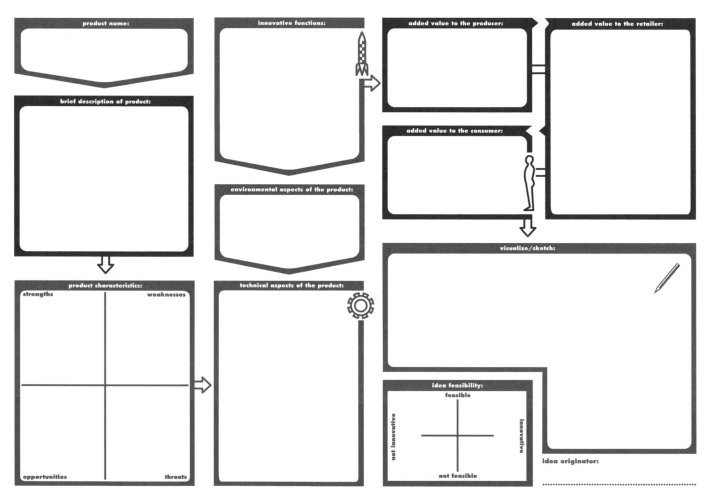

If the idea process end-goal is some form of project or action plan, it may well be appropriate to apply this graphic design to advance the ideas to the realization phase. Collect the concepts that have been enriched and that have undergone the final screening and prepare them for realization using this tool.

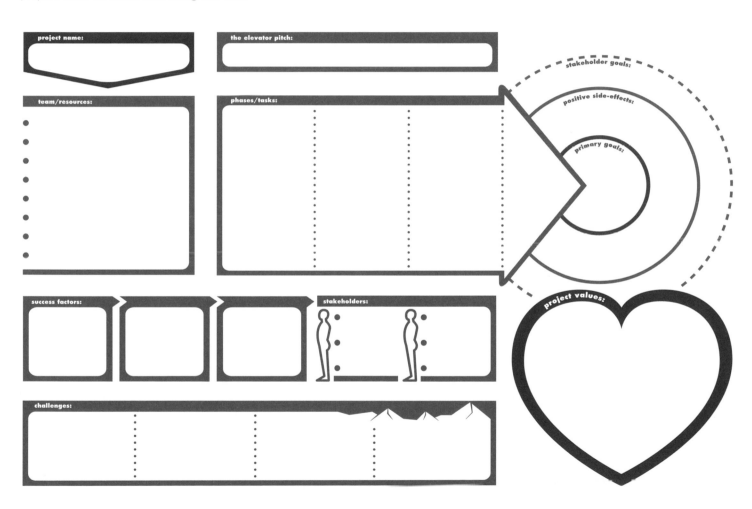

chapter 8
RESULTS
– and what next?

OK then – now you have a whole bunch of concepts that you think are brilliant and you're just waiting for the world to pat you on the back. But sadly, it's rarely that simple. Your ideas now have to be made presentable – you need to convince the decision makers of their finer qualities, and your concept must be championed with the help of your own creative courage and that of your team. The warrior, the worker and the visionary within you must all collaborate to promote your world-beating concept.

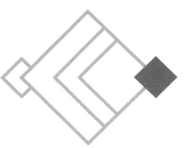

WHY DOES NO ONE APPRECIATE MY BRILLIANT IDEA?

But presenting your idea, underpinning it and captivating people with it is not an easy task. Just ask Columbus, Einstein or even Bill Gates – they all had fantastic ideas with incredible potential and were initially told that they were more or less nuts. It can be worth reflecting on the reasons that people resist innovation before you try to appeal to the innate conservatism of human nature:

- They don't want to change their culture. Humans are creatures of habit and entire societies are founded on law and order, frameworks and procedures. To start with, it's much easier both personally and socially to "keep doing things the way they've always been done."
- They don't understand the idea. Good communication is the hardest art to master and it can take several presentations before you can get your message across. People are also scared of asking dumb questions that make them look foolish and would rather keep their ignorance quiet.
- They don't like the person presenting the idea. Make sure that at least some of the people that you're presenting ideas to are, if not allies, then at least positive to you and your persona.
- They don't think the idea is any good and/or they don't believe that it will solve the "real issue." This shouldn't happen if you've carried out the need orientation correctly.
- They can't see how the idea can be implemented, or whether it will need a greater

work commitment from each individual. In the absence of precise definitions people always assume the worst.

■ They don't understand how their own role will change if the idea is carried through. Will they lose the power and/or respect that they've established over the years as a result of the new idea? Will there still be room for each individual?

■ They can't see what's in it for them. How will they benefit if the idea is implemented? If there's no obvious gain for the individual in question, then the incentive to listen may well be lost.

In addition to giving inspiring, captivating and visual presentations, it can be an advantage to think through the above points with a view to reassuring your audience by providing them with an element of security and confidence. Documenting parts of the creative process on video and then illustrating the energy and enthusiasm of the creative team at work can be a constructive complement to the results phase. There are no easy solutions to this problem, but you should design a strategy that will help you resolve similar situations if they develop in your organization. Try to ensure that decision makers are included as "part of the solution" and make them feel that it's as much their idea as an innovation developed by you and your creative team. Then release the "idea virus" and try to convince the entire organization to accept and embrace your idea!

YOU'RE NOT ALONE

Sadly, the romantic buzz of the initial stages of the idea process does not last indefinitely. Working with innovation is a continuous roller-coaster ride that often requires a great deal of determination. The realization of ideas is an uphill struggle of real-life challenges. If you encounter one of the following 10 setbacks, you should know that you were not the first…

1. At the end of a creative session, someone gathers up the ideas and notes for documentation. Three weeks later, when you get the chance to look at the printout, you don't feel the same excitement as you did when the ideas were first generated. In fact quite the opposite: you wonder whether these notes are from the same session at all.

2. Midway into a project, the "current market situation" results in you only being given a fraction of the resources that you were promised to carry out the project.

3. Your technical colleagues shake their heads and say: "Sorry pal, but we've changed our minds. We can't do it to the original specifications."

4. Your project never seems to reach the top of the priority list.

5. Your new Director of Finance wants to know exactly how much revenue your idea will generate.

6. The Board suffers collective memory loss and forgets why the idea was so brilliant in the first place.

7. Your colleagues stop calling you with spontaneous development solutions that would have become your next step.

8. Your boss is transferred.

9. You hear that the competition is about to launch a similar product.

10. The idea that you were pushing and that didn't get any response shows up in someone else's presentation. All of a sudden, it's a fantastic idea and your colleague gets all the credit for it.

After all the items in this list have been ticked off and you feel you're on the verge of giving up, just remember – you are not alone, so keep fighting! Both you and those around you will feel much better if you do.

THE COURAGE TO CREATE

For many people nowadays, the most important thing in life is daring to live one's ideas and dreams, and achieving self-fulfillment. To have a dream, an idea and realize it – TO CREATE. But if creating is the most fulfilling experience a human can enjoy, it is also the part of the life process that requires most courage. Why? Because it takes guts to find new approaches, to explore uncharted territory and to put your reputation on the line without knowing for certain if you're right to challenge conventional wisdoms. Setting off into unexplored terrain can prove a rocky experience with a great many pitfalls. And it is only the creators themselves that can even see this fantastic world that lies at everyone's feet. The courage needed to achieve this can be related directly to the implementation required. The greater the challenge and the more revolutionary an idea, the more determination it needs to realize it. This is why it's essential that a creative hero, an entrepreneur or an artist has an innate belief in him or herself, trusts in his/her emotional instinct and ideas, and understands that the process of change often involves overcoming challenges and conflict.

If you dare go the whole hog, most evidence indicates that you'll live a happier and more fulfilling life, and that you'll be enjoying a privilege that, according to James Joyce, only visionaries and audacious creators are granted – that of helping to "forge the uncreated conscience" of the human race.

METHODS

NEED ORIENTATION

IDEA GENERATION

SCREENING AND DEVELOPMENT

INTRODUCTION AND ENERGY TOOLS

5 WHYS?

5 Whys? is a practical method that can be applied during an initial discussion with a customer when defining the focus area. The technique was pioneered by Sakichi Toyoda and involves simulating the behavior of a five-year-old child by continually asking questions until a true understanding of the central issue is achieved: "Explain the problem to me as if I were a five year old." 5 Whys? is in reality more of a mental tool than a method, the point of which is to alternate between extremes of abstraction and tangibility to reach a formula that is neither too wide nor too narrow. The method is also useful in bringing an issue into focus by approaching it from different levels with a view to identifying its true nature.

Step by step

1. The need is discussed and defined on as tangible a basis as possible. Make sure that it's simple and clear.

For example: The customer is having trouble hiring new staff.

2. It's then made increasingly abstract but more profound by asking the question: Why?

For example: Why do we have a problem hiring staff?

3. The answer to this question will often form the basis of a focus area that can lead you to the solution.

For example: The customer's brand image is not attractive enough.

The resulting question would then be: How can we make our company more attractive to prospective employees?

4. If you don't think your question has succeeded in reaching the heart of the issue, keep asking the question "Why?" until you do:

Why are we not sufficiently attractive to prospective employees?

But try not to apply excessively abstract formulations. These may provide some perspective on the focus area, but they won't produce a satisfactory result.

NEED ORIENTATION TOOLS
ways of identifyng the root cause
and the real need

SITUATION

5 Whys? is very effective when the issue/question is blurred or when a deeper understanding of the root cause is required. A mental tool such as this is especially relevant as a means of generating a discussion around the central issue. The best focus is often a combination of tangible approach and visionary thinking. The method is dependent on a team leader who is skilful at identifying underlying patterns in issues. It is specifically designed to help people who have difficulty interpreting their own situation or approaching a problem from a broad range of perspectives.

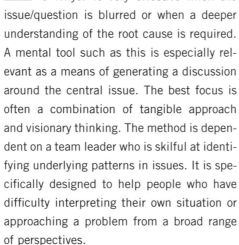 **RESULT**

A clearer understanding of the "real" need or root cause, which usually involves formulating or defining the issue.

 PARTICIPANTS
1–8.

 TIME FRAME
5–30 minutes.

 ENVIRONMENT
No specific environment is needed.

MATERIALS
Whiteboard.

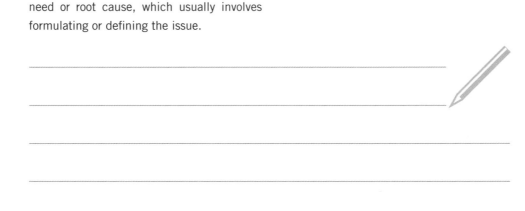

PRIORITIZATION ANALYSIS

Finding a solution to the central issue is the essence of all forms of idea development. To achieve this, it can sometimes be useful to carry out a prioritization analysis, either individually or with your creative team. Make sure that you're analyzing causes and not symptoms. Based on your own thoughts and the input of the rest of your team, what is the central issue and what are your options?

Step by step

1. Write down what you think is the goal of your task.

For example: Make the US into the world leader in ideas.

2. Ask yourself: Why do I want to achieve this task? Write down your purpose.

For example: I want to pass on my knowledge to others.

3. Ask yourself: What is my biggest challenge in achieving this task? Write down your answer.

For example: I have to put my thoughts down in writing and try to communicate them to my colleagues.

4. Read through your three answers again and reflect on them. Steps 1–4 should take about 30 minutes.

5. If you're working with your team, add another 15 minutes so you can compare answers.

6. Then reason your way through to the central issue or your real options, and define a focus area.

SITUATION

When you want to analyze your impression of a situation as well as your own driving forces, then compare them with your team members to build a sound base for additional idea creation. Using this method, you can analyze the situation from your own perspective, while at the same time trying to understand the individual driving forces that are essential in defining the focus area. Bear in mind that this exercise is primarily for mapping an internal situation that forms a basis for defining the focus area. This method is useful when there are a large number of different thoughts and opinions in circulation.

RESULT

A clearer picture of the need. The team can actively air their thoughts and views. Provides a feeling of consensus around the establishment of the focus area.

PARTICIPANTS

Everyone participating in the idea process, not only the project leaders.

TIME FRAME

30 minutes for individual work, 60 minutes in a group.

ENVIRONMENT

Anywhere.

MATERIALS

Paper and pens.

RADICAL HYPOTHESES

Radical innovation has become a buzz term in recent years. Put simply, it describes those paradigm shifting solutions that completely alter an industry, such as Spotify's distribution and payment solutions in the music industry or Skype's technical solutions in the world of communication. If you are consciously looking for this type of revolutionary idea, then it can be helpful if you are already in the need phase of work with "unthinkable" angles and different hypotheses. This method helps you discover the traditional truths and norms in your industry in order to find new opportunities for big ideas. It is more looking into opportunities than digging deeper into a problem. The American Luke Williams from the design firm Frog Design inspired this approach with his book *Disrupt* (2010). His favorite example is how someone in dinner conversation came up with an idea to start a business selling socks that did not match. Everyone thought it was a lousy idea. A few years later the company Little Miss Matched became a great success with their different colored socks sold in "pairs" of three.

Step by step

1. Write down the the focus area you want to create new solutions within. It could be your industry or a specific product, and it helps if it is an area which seems to have stalled for some time and needs to be developed.

Example: How can we renew the package and delivery industry (both products and services)?

2. Ask yourself or your team: What are the obvious clichés in this area?

Example:

A. All packages and letters look the same.

B. Packages and letters always seems to be delivered by men.

C. The uniforms are ugly.

D. Those who deliver are perceived as stressful.

3. Think about which three scenarios you can create around every cliché. Try to put things in a different perspective and do not be afraid to provoke, if you can – it often improves the result.

Example (Option A in this brief example):

Imagine if all the packages and letters were unique... Imagine if all the packages and letters could tell where they were heading and could tell a story about where they were going and how ... Imagine if all the packages and letters smelled different depending on who sent them...

4. You should now have a number of hypotheses to choose from. Try to find a common thread and describe some brief insights and opportunities around the most interesting hypotheses.

Example – insight:

The outside of mail and packages often do not engage people. They are convenient but boring and have very little soul and smart features. And they are delivered in an impersonal way.

Example – opportunity:

Can we develop a company or a service that creates personalized, interesting and clever packaging delivered by personnel who creates experiences for the recipient?

SITUATION

When you want to explore opportunities for entirely new approaches and ideas in an industry that has been standing still for a while. Clichés and hypotheses will help you turn things on their heads even in the need phase.

RESULT

A number of exciting new opportunity-based needs for developing new ideas in a stagnant niche or industry. The method has a clear humorous touch, and gives rise to both laughter and new insights.

PARTICIPANTS

5–20 persons – groups can be both external and internal.

TIME FRAME

15 minutes to list a number of clichés. 30 minutes to create hypotheses/clichés, preferably done in small groups. Finish with a 30-minute plenary discussion and define an insight and an opportunity for the idea generation phase that follows.

ENVIRONMENT

A room with good wall surface or whiteboards to collect and present the final results of the discussion.

MATERIALS

Paper and pencils. Preferably colored paper for the different steps, so that it is easy to see what's what.

THE RELATION MATRIX

The aim of this method is to create a better understanding of the issue as well as quickly to identify underlying causes two or three levels below it. It is inspired by the Japanese quality movement Kaizen, best known for its continuous improvements at Toyota, and is an excellent exercise for really reaching to the core of a problem, the root cause. You have probably – as many of us have – been faced with the frustration of working hard for a solution only to find that you've solved the wrong issue. The Relation Matrix ensures that you achieve a true understanding of the central issue. The Idea Agent is relatively passive during this exercise – giving instructions, coaching the creative team during the process and managing the schedule.

Step by step

1. Introduce the focus area, which doesn't need to be defined in great detail.

For example: How can we become more creative in our organization?

2. Team members begin by writing down individually sub-issues or obstacles on Post-its that they think are relevant to the focus area.

For example: What is preventing our organization from becoming more creative? Put your answers up randomly on a whiteboard.

3. Then let each team member take down a Post-it from the whiteboard and read what it says to the rest of the team. Make sure that everybody understands the sub-issue! There is no need for consensus but the issues must be understood and discussed during this part of the process. As an Idea Agent, you facilitate this discussion and ask relevant questions so that there is no doubt about the meaning of the message. They should then place the Post-it on a second empty whiteboard when they've finished. The next team member repeats the process but also tries to find points in common with previous Post-its. If the notes have similar themes or subject areas, place them next to each other on the board and allocate a title to the cluster. When all the notes have been read and sorted, the team then discusses whether respective cluster titles are accurate.

4. To develop a deeper awareness of context, arrows can be drawn to indicate the relationships between the clusters.

5. Finally, each team member allocates five points or dashes to establish which sub-issue should be given the highest priority.

If the original team was quite large and split into two groups, you should now encourage them change places and continue the process with the other group's results. This is done to generate new energy and so that team members can find out how others have perceived the same issue.

SITUATION
When the focus area/issue is blurred or when you want to increase awareness of the sub-issues.

RESULT
A detailed description of the central issue and a corresponding description of solutions to all the identified sub-issues. If you then want to develop one of the sub-issues, you can do this using an idea generation method.

PARTICIPANTS
6–10 per group – make two groups and you might like to encourage the groups to change places midway through the exercise.

TIME FRAME
2 minutes of introduction + 30–90 minutes of issue orientation.

ENVIRONMENT
A room with lots of wall space or even separate, movable boards to work on. Make sure to place chairs nearby so you can sit in the background briefly and watch the groups at work.

MATERIALS
Post-its and colored pens.

RANDOM WORD ASSOCIATION

The ability to associate is fundamental in idea generation contexts, for example contrasting images will spring to mind for each of us when we hear the word "lawnmower." These could include warm summer days, the smell of new mown grass, a chore that just has to be done, or a distant childhood memory. The diversity and power that are intrinsic to a group's collective associations is what we are attempting to tap into with this method. Random Word Association can be applied very effectively when a large number of ideas have to be generated in a short space of time. Several hundred is not an unusual yield.

Step by step

1. Describe the exercise one step at a time. Wait until step 4 before you introduce the issue.

2. Ask one of the team members to start by suggesting a word. This word can be anything, something they can see in the room or something in their minds at that particular moment. In our example we've used the word "lamp." Write up the word on a whiteboard and ask the next team member what associations they can make with the word.

3. Do this for about five minutes to create a word association chain. For example, this could be: lamp, light, summer, vacation, London, rain, umbrella, etc.

lamp	lightspeed	love
light	blink	rose
buoy	glasses	red
sea	cobra	blood
starfish	bite	accident
space	venom	hospital

4. When the chain is "finished," team members have warmed up a bit, and they've grasped how associations should be made, introduce them to the central issue. The reason that this should be done at this point is to avoid the issue interfering with the

chain. If the issue is introduced too early, the words will almost exclusively center around the issue and fewer innovative solutions will be generated during the idea generation phase.

5. The time has come to begin manufacturing ideas. Make sure that each team member has a pen and a Post-it pad. The Idea Agent then reads out the issue and relates it separately to each word in the chain.

For example: "What can we do to prevent a future teacher shortage?" in relation to the word "vacation"?

Now it's no holds barred! The task of the team is to associate freely around solutions to a teacher shortage in relation to the word "vacation." Each team member expresses their idea out loud to inspire the others and then promptly writes it down on a Post-it.

The others will hopefully make an association with the previous solution or idea, say it out loud then write it down. Sometimes it can be a good idea to motivate team members with questions during the process.

For example: a team member declares: "We should offer teacher-led theme trips." Ask them: "What do you mean? How would they be organized?" This way you can "force" team members to be more specific with their descriptions and enrich their original ideas.

When the creative energy has dried up – pick a new word.

6. Finish by allowing team members about 10 minutes to develop their own ideas so that the end result is not a bunch of notes with only one word on them. The ideas will have much more value if there's more substance to their descriptions.

SITUATION

When it's important to generate lots of out-of-the-box ideas. Random Word Association is very effective for finding brand names, for example.

RESULT

A pool of ideas that might seem a little too underdeveloped to work with, but which are often usable in the end. This exercise will produce a lot of "not-thought-of-before" ideas.

PARTICIPANTS

Between 6 and 10. People with a gift for ideas generally find this method effective, but there is also room for a more thoughtful approach.

TIME FRAME

About 40 minutes.

ENVIRONMENT

Set up a table in front of a whiteboard and put chairs on three sides of it so that everyone can see the whiteboard.

MATERIALS

A table with chairs placed so that everyone can see the whiteboard and the Idea Agent. Post-its and pens.

FORCED COMBINATIONS

Finding a winning concept sometimes involves creating new recipes with the help of old ingredients. Though fresh foreign culinary influences can appear now and then, most food is cooked using ingredients that have been available for the past 50 years at least. And yet newly composed dishes are constantly appearing on the menu of any self-respecting restaurant. Forced Combinations is a method that defines the existing features or components of a product or service then tries to combine them in a new way. It can also be applied during the idea development phase if you find the ideas that you've generated are too fragmented. Forced Combinations is an exercise that originated in the automotive industry. It's a very effective method primarily in product development processes, but can also be used to good effect in other contexts. Achieving a satisfactory outcome requires a good deal of thought and planning.

Step by step

1. Identify the essential variables or features that an end product might include.

For example: If you want to target a new food product at a particular a youth segment, the variables might be ingredients, design, taste, image, packaging, etc.

You can determine the essential variables before the process begins, or you can do it together with your team.

2. Write up the variables on a board in a long row so that everyone in the team can see them. Then ask each team member to write down variables on a sheet of paper as well (if you haven't made templates in advance).

3. Now ask the team to think of different forms, types or features for each variable and write them down under the variable that they relate to. Make sure you get at least 15–20 but not more than 30 practical suggestions in each column.

For example: Different types of ingredient could be milk, meat or vegetables and different forms of identity might be rapper, jock, computer nerd etc.

4. Team members should now work individually to combine the types or features from each column into a new concept based on these components. It's important to write down the combination of features that will form the new concept so that you can explain and inspire others at a later stage in the process.

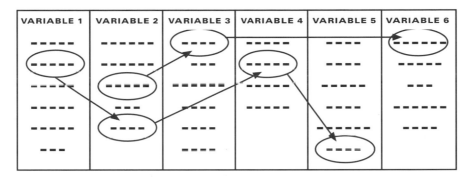

| VARIABLE 1 | VARIABLE 2 | VARIABLE 3 | VARIABLE 4 | VARIABLE 5 | VARIABLE 6 |

For example: The mix of features "meat – round – spicy – cool – bottle" could become a spicy Tex-Mex meat dish in a bottle called Cool Pedro.

5. When energy levels are starting to fade, encourage team members to interact with each other – for example one-on-one – to enable new ideas to develop or even generate a few new concepts.

 SITUATION

Forced Combinations is primarily suited to the development of an existing service or product, but can also be applied very effectively in the development of step-based processes. Your team doesn't need to be highly trained in creativity methodology to use it.

RESULT

Well-developed ideas that are usually realizable. Not to mention all those possible combinations – seven variables x 20 ideas per variable equals 7^{20} ideas.

PARTICIPANTS

5–15.

TIME FRAME

A good deal of preparation time is needed to find the right variables to work with. The exercise itself can last from 40 minutes up to 2–3 hours depending on how far you want to develop each variable.

ENVIRONMENT

Try to use a large room with lots of wall space. Play background music during the individual phases.

MATERIALS

Paper, pens and a large board to write up variables and features.

NEGATIVE IDEA GENERATION

Turning commonly accepted wisdoms on their heads is one of the most powerful tools in creative management. One approach to viewing something from a new angle can be to reverse your perspective from the positive to the negative. The principle of Negative Idea Generation is that sometimes it's both easier and more fun to knock down, beat up and backbite than it is to build up. Put simply, some people find it easier to generate negative ideas – ideas that are a bit more frivolous and work against the grain – than positive ones. Negative Idea Generation is also an excellent method for idea generation in large groups. You should play high-energy music in the background during this exercise.

Step by step

1. Define a positive focus area that's fun and inspiring and that you can "turn on its head" and into a negative.

For example: How can we get baby boomers to invest in newly built condominiums one sunny day 10 years from now?

2. Split your team into groups of six to eight around each table so that the groups can see each other. If there are several groups, you can have different focus areas without disturbing the process.

3. Explain the exercise to the groups. Tell each group to "reverse" the positive focus area for about five minutes and create a negative-sounding focus area. This can be done in various ways, but it's important to retain the essential elements of the original focus area to achieve a satisfactory result.

For example: How can we get baby boomers not to invest in our real estate one rainy Monday 10 years from now?

Then give each team member a sheet of A4 paper – either a grid like the one on the next page or plain white, and ask them to draw a large cross on it. In the top left-hand corner they should indicate the positive focus area and in the bottom right the negative one.

4. Now encourage the groups to generate ideas for about 15 minutes based on the negative focus area that they've created. The more off-the-wall the ideas are the better – tell team members to really make the most of their chance to play devil's advocate.

Ask team members to write down their solutions to the negative issue in the field in the bottom right of the page.

5. You should wind up the generation of negative solutions when an acceptable number of ideas has been generated. Now take about 15 minutes to transform the negative

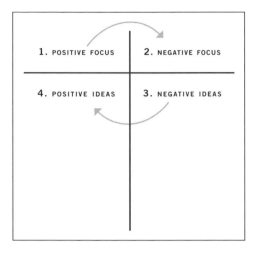

| 1. POSITIVE FOCUS | 2. NEGATIVE FOCUS |
| 4. POSITIVE IDEAS | 3. NEGATIVE IDEAS |

ideas into positive equivalents that relate to the positive focus area. It should be quite easy to reverse some of the negative ideas directly into positives while others will generate a whole new range of possibilities. Ask your team not to make their transformations from negatives to positives too simple! Then make sure that they write their ideas in the field in the bottom left of the page.

SITUATION
Negative Idea Generation is useful when working in groups of over 10 people. You can manage a process with a very large number of participants by splitting them into groups of 6–8 people. The method is also practical when the team includes a large number of negative personalities. Allow them to say "no" for about half an hour and the climate will improve.

RESULT
Lots of fun and high spirits, but also ideas that range from the excessively logical to the really wild.

PARTICIPANTS
6–100.

TIME FRAME
30–60 minutes.

ENVIRONMENT

A large room with separate tables seating 6–8 people.

MATERIALS

Paper and pens. Use a template during the process like the one in the illustration on the previous page.

WHAT IF?

What If? is a method that uses contrasting roles, professions, situations and brands to assist in examining and solving central issues. It involves trying to "play" different people or things, then trying to imagine how these people would have solved a particular problem. It's a method that is reminiscent of role-play. How much you choose to take center stage is up to you. But for What if? to succeed, you will need to do a good deal of preparation. Determine which roles are most appropriate for a particular context. Which customer groups are you focusing on? This method can also be applied in real life if there is a need. The design firm IDEO usually sends its designers on location, for example to a department store, to enable them to study and create ideas on the basis of consumer behavior in a particular place and with a particular product.

Step by step

1. Prepare a scenario for a number of appropriate "roles" or customer groups. Think through how they can be presented as realistically as possible.

2. Present the central issue and then one "role" at a time. Ask team members to write down their ideas on Post-its.

For example: The case is about designing a new handbag. The first role could be: an elderly lady who is coming down the street with a walking stick in one hand and a grocery bag in the other – which solution would she appreciate in relation to the central issue? Which elements would be important to her? A young basketball player is chilling with his homies as he shoots a few hoops and drinks soda – how would he want our product to be designed?

3. If the "role" that's intended to inspire doesn't result in more ideas, you should use a new one from your list. This list should be drawn up and designed to create the optimal dynamics around the central issue. There is often a good balance if you choose a few "roles" that are close to the issue and a few that are more far fetched.

4. Wind up the idea generation process and ask your team to write down their ideas. Make sure the ideas are clear and add a few details so that they'll be understood at a later stage in the process. Give team members five to 10 minutes to do this and play a little music.

SITUATION
When you have a clearly defined target group for your idea generation.

RESULT
Both grounded and inspirational ideas.

PARTICIPANTS
6–12.

TIME FRAME
30–45 minutes.

ENVIRONMENT
A nice table where everybody can see each other.

MATERIALS
Post-its, pens and a well-thought-out list of roles and situations.

THE DREAM TRIP

Dreaming and visualizing can be a good technique for generating new ideas. The power of imagination is a very useful means of breaking free from the inhibitions in which common sense and habit have often chained us. The Dream Trip utilizes this to help create ideas and to encourage a visionary climate in the team. The method is best used in the initial phases of idea generation when there is a broad or loosely defined focus area.

Step by step

1. Make sure you prepare the Dream Trip thoroughly before you start. The method can be based on several scenarios with different ulterior motives. Choose one from these three examples that suits your purpose best.

a. Imagine that you're standing somewhere in the future reading a newspaper. There's a report on the front page with a large photo describing how you managed to find the perfect solution to the focus area and what the result was. As you're reading the report, you notice that everything you dreamed about and that you thought was virtually impossible has been achieved. What is it that you created? How exactly did you create it?

b. Imagine that you're taking a trip to a place in the future unrelated to the focus area. This could be a trip to the zoo; you could be catapulted into an episode of *Star Wars*, or just browsing round a big city. Then let your team "introduce" the focus area into your dream scenario and see what solutions develop.

c. Take a trip to the future and a place where the solution to your issue already exists. What can you see around you? Who's there and what are they doing? What feelings do the images you see generate? How do events unfold? What are the solutions in this world?

Try to dream your way through the imaginative journey and then write down a few key words from your story.

2. Introduce the focus area and answer any questions. Then ask team members to shut their eyes and listen to the start of the Dream Trip. Begin describing the scenario that team members find themselves in. Talk slowly and pause regularly so that they have time to form their own images and visualize their own fantasy world. If you have music that you think might suit the mood, play it.

3. When team members have been visualizing for about five to 10 minutes, ask them to open their eyes, reconnect with the focus area and write down the ideas that they envisioned.

4. When they've written down their ideas, they should describe their experiences to each other – in small groups or as a team. It's essential to allow them the space to explore each other's thoughts so that they can develop new ideas from the visions that are being described. Ask them to enrich each other's ideas and to generate new ones as they take turns to present, and remind them to write everything down.

5. Allow team members some alone time after this process to generate new ideas or enrich the ones that they heard during the exercise.

 SITUATION
When you're looking for ideas "on a higher plane" or if you want the team to develop broader visions.

RESULT
Visionary ideas that sometimes need a short process of individual enrichment.

 PARTICIPANTS
Most suitable for team members that can visualize individually. The method can be used with up to 15 people.

TIME FRAME
45 minutes.

ENVIRONMENT
In an inspirational room. Accessorize the room to suit the selected Dream Trip and enhance the experience.

MATERIALS
Paper and pens.

DAY PARTING

Does time of day change what your needs are and what products or services you may need? Of course. An attractive way to work with the development of ideas is to try to understand the customer's days, and his or her activities, to generate ideas. The foundation of the method is a simple analysis of a person's 24-hour schedule. It begins with a worksheet or whiteboard with a timeline that goes from 00.00 to 24.00. In the example below, we use media consumption as a subject area, because media today is certainly something that is consumed around the clock. If the product or industry is a little more limited in time or behavior, it is perfectly fine to have a shorter timeline or a timeline that is also tied to a geographic location.

Step by step

1. Draw a horizontal timeline with start 00.00 and end 24.00 on a large sheet or wall.

Example: The issue is: "How can we expose our brand to our target group at different times of the day?"

2. Write down a number of assumptions about what the customer is doing at various times of the day.

Example: 06.30 Anna wakes up and wakes up the kids. 7:00 Anna fixes breakfast and speaks on Skype with her husband who is on a business trip. 7:30 Anna takes the car to work. The kids are off to school.

3. Add another layer that connects different types of media and media channels to the customer's activities throughout the day.

Example: 06:30 The alarm on the smartphone that Anna uses as an alarm clock, goes off. The smartphone displays time, the weather report and the latest news as she picks it up.

07:00 At breakfast the TV is on in the background. On the iPad, that is lying on the dining table, is Facebook, email and nytimes.com on the screen.

7:30 In the car, Anna listens to the local radio station to find out if there are any traffic jams. She also consults her GPS in order to find the fastest way, based on morning traffic.

4. Add a final layer that deals with the various interaction opportunities for the brand or the product you are working with. Create many concrete idea on each occasion describing what this interaction could look like.

Example: 06.30 Can we help Anna in some way even as she is waking up. Maybe a mobile application that exposes our brand?

07:00 Anna is on Facebook, how can we interact with her there in the morning? What solutions could we provide? What can we sell to her there? Or should we skip the hard sell in the morning and just warm up with a little bit of fun brand interaction?

07:30 Can we send out relevant brand messages through advertising on the local radio station or through geo-location-based services? How can we help and interact with Anna in the morning traffic?

 SITUATION
When you want to get an overall picture of a customer's behavioral patterns. Suitable for examples of issues related to marketing or consumer-related industries.

 RESULT
A number of "maps" that describe some customer-types' behaviors, over the course of a day. Creates numerous and concrete ideas.

 PARTICIPANTS
6–30 participants in groups of 2–4 people.

 TIME FRAME
60 minutes – spend 15 minutes each on the first two steps and 30 minutes on the last phase of idea generating.

ENVIRONMENT
Plenty of room on tables or walls to enable the participants to work with their timelines.

MATERIALS
Pens and large sheets of paper that can be put up on a wall in order to easily produce and present each group's timeline.

VISUAL CONFETTI

Some people are inspired by words and verbal forms of expression, others by images, colors and shapes. Visual Confetti is an associative method that applies the visual in creating new ideas. It involves compiling a collection of images such as spectacular abstract motifs, or photos of objects with a relation to the focus area.

There are quite a few variations on this exercise. Your images don't need to be projected on a wall and can be used in smaller sizes if more convenient. You can show them in bunches, or individually as in the example below. You can also develop the method by asking your team to describe the smells, sounds and other impressions that they associate with the images, and this information along with the images themselves will enable them to generate ideas.

Step by step

1. Compile about 30 images that you think are inspiring or relevant to the focus area. They could be of the Great Wall of China, a Van Gogh painting, a spider monkey in the Amazon jungle or even a competitor's product.

2. Ask your team to sit in front of you in a semi-circle and make sure that they all have a pile of Post-its and a pen. Introduce the focus area. Stand next to the image projected, facing the team, so that you can talk to them while they're viewing the images.

3. Show the first image then read out the focus area and ask the team to look at the image for inspiration.

> For example: If the focus area is "How can we get Generation X to buy our white goods?", the Idea Agent should say: "In relation to this image: How can we get Generation X to buy our white goods?"

Anyone who has an idea should say it out loud and then write it down on a Post-it. And one idea per Post-it is the rule. It's important that you coach the team by continually spurring them on with praise and encouragement. As soon as you notice that they're getting a bit passive, it's time to change image.

4. When an acceptable number of ideas has been generated or energy levels are flagging, you should wind up the exercise.

SITUATION

Visual Confetti is designed for people who think in images. The method is useful when alternated with exercises that are more verbally based.

RESULT

Tangible ideas with a wide range that can be enriched both in written and graphic form.

PARTICIPANTS

1–800.

TIME FRAME

45 minutes.

ENVIRONMENT

As bright as possible, but not so bright that the team can't make out the projections.

MATERIALS

A projector, images in digital or physical form, pens and paper. Allow team members to document their ideas as both images and text.

HEADLINE MANIA

Headline Mania is a method that applies the media expression "sampling" to idea creation. Headlines often contain simple, expressive words and project powerful messages. The purpose of this method is to recreate the atmosphere of curiosity and learning that characterizes a library reading room.

Step by step

1. Purchase a few newspapers and magazines that are either completely unrelated to the focus area subject-wise or that seem interesting from a future or trends perspective.

2. Consider the focus area and define it as a question.

3. Give each team member a few newspapers and magazines and ask them to find a few headlines or themes that might be interesting to develop.

4. Then focus on one theme or headline at a time and test it in relation to the focus area:

For example: How can we solve (the focus area...) in relation to (one of the selected headlines or themes...)?

5. Each team member then says their ideas out loud – which will help inspire the others – and writes it down straight away on a Post-it. One idea per Post-it is the rule, and ask team members to write as clearly as possible.

6. Take up each theme or headline for a few minutes until energy runs out. Then take turns to choose themes.

7. Finish the exercise by sitting individually and developing the idea fragments that have been generated.

8. Put all the Post-its up on a wall for a good overview.

SITUATION
When you need ideas grounded in current tendencies.

RESULT
A large number of tangible ideas. The range of results is dependent on the selection of newspapers and magazines.

PARTICIPANTS
Up to 15.

TIME FRAME
10 minutes of introduction + about 30 minutes in groups + 5 minutes individually.

ENVIRONMENT
Use comfortable chairs and try to recreate a library atmosphere.

MATERIALS
Pens, Post-its and a range of interesting and inspiring newspapers, journals and magazines.

6-3-5

6-3-5 is a method for individual idea generation. The method is normally used with a team that has a good awareness of the focus area and may even have had time to consider the central issue. The name of this exercise is derived from the fact that six people should produce three ideas every five minutes. Theoretically, this would mean an idea generation yield of 108 ideas in 30 minutes!

Step by step

1. Ask the team to sit around a table so that they are within arm's length of each other on either side. Introduce the focus area and make sure that you write it up on a board so that all the team members can see it.

2. If the Idea Agent has prepared templates, these should be passed around – one worksheet per person. If there are no prepared templates available, you should explain to the team how each sheet of paper should be drawn up. Then hand out blank sheets – preferably A3 – and ask your team to draw a matrix on it with three columns and six rows (or the number of team members participating at the time).

3. Then for a short time – say five minutes – each team member should write three ideas in the three top fields. After the first round, you can either develop the ideas as they are or let them inspire you to a completely new idea.

4. Pass the papers around the table. The task of the next team member is to generate three new ideas based on the ones in the top three fields. Keep going until the papers have gone once round the table.

SITUATION
If you're short of time and need to generate a large number of ideas.

RESULT
A large number of ideas, some of which will be relatively well developed.

PARTICIPANTS
5–8.

TIME FRAME
30–40 minutes.

ENVIRONMENT
Preferably a round table.

MATERIAL
Pens and paper – preferably A3. Prepare some kind of template to enable the idea process. If you don't have the time to do this, be careful to demonstrate how the columns should be drawn and how to write down the ideas. You don't want your team to worry that they're doing things the wrong way instead of applying all their creative energies to generating new ideas.

FISHING STORIES

Exaggeration is a means for us humans to apply our imaginations to differentiate between the best, the worst and the biggest. And most of us remember those exaggerated stories about the bass in the local lake or the touchdown they scored with their high school football team, right? The creativity required in describing these amazing experiences is hard to miss. If you take time to listen to how children fantasize, you'll notice that they often use various types of exciting and vivid exaggeration. Fishing Stories is grounded on the ability to exaggerate to create new and exciting ideas.

Step by step

1. Prepare a number of exaggerations that might be considered relevant to the focus area.

2. Start by describing the central issue or the need.

3. You should then describe a series of exaggerations – at least 10 but preferably more – to stimulate a few wacky ideas. Introduce them one at a time and give the team a moment to think of ideas for each exaggeration. You can use this example before introducing exaggerations so that your team gets the picture.

> *For example: The subject is product development for football players. Exaggerations might then be:*
> *– They're all as rich as Bill Gates.*
> *– They all have pop star girlfriends.*
> *– None of them has ever been to the opera.*
> *– They all think they're misunderstood geniuses.*

4. Write down all the ideas that come up or ask your team to write them down.

5. Use these ideas as a springboard to new tangible concepts and ideas. Then let the team sit individually for 10 minutes to enrich them.

SITUATION

There will often be a product as starting point, for example something that has to be tailored to a new market.

RESULT

Lots of laughter and crazy ideas. When the ideas are confronted with reality, new perspectives and opportunities will usually develop.

PARTICIPANTS

5–12.

TIME FRAME

About 45 minutes.

ENVIRONMENT

Around a table.

MATERIALS

Materials for illustrating your exaggerations. Paper and pens.

IDEA PROPPING

The name Idea Propping implies that ideas are created using props, a prop being any movable articles or objects on the set of a play or a movie such as the sledge in *Citizen Kane* or the ice pick in *Basic Instinct*. The method involves each team member visiting different prop stations to look, feel, smell, taste, listen and, not least, generate solutions to the defined issue. Your team should feel inspired by and be able to associate around the objects that have been placed at the different stations. This method is primarily an individual journey.

Step by step

1. Select a mixture of objects that you think are both close or irrelevant to the focus area, or whose originality will prove inspirational. Remember that inspiration can derive from the feel of an object, its taste, symbolism, applications etc. Your chosen objects can be anything from an apple to a dish brush or the latest cell phone. Think about the type of associations you want to generate in the minds of your team. Try to include as many senses as possible.

2. Set up a series of stations containing one object each. Each station should also include a Post-it pad and a pen.

3. Make sure that each team member takes their place by a station – but only one person per station. Now they should start generating ideas inspired by the object that they see in front of them. Each idea should be written down on a Post-it and applied in the space around the object. After about three minutes, you should indicate that it's time for a place change upon which everyone should move one place clockwise and continue the idea generation process with the help of the new object and the ideas that are already there.

4. When every team member has visited each of stations, allow three "bonus rounds" during which team members can visit their favorite stations and generate new ideas in discussion with the people who are already there. But they should shift places at every place change and visit three different stations.

5. The ideas are then collected up and put on a wall for further development.

SITUATION
Idea Propping is a method that enables individual and reflective idea generation. It's a relatively simple method to manage, but requires a good deal of planning time since the quality of the various stations has a direct impact on results.

RESULT
A large number of tangible ideas whose direction the Idea Agent can influence through his or her choice of object.

PARTICIPANTS
5 or more.

TIME FRAME
About 45 minutes.

ENVIRONMENT
Plenty of space so that the different stations and seats can be established.

MATERIALS
Pens, Post-its and a selection of physical objects that are close or unrelated to the focus area or whose originality will inspire.

MERLIN

Merlin was the wizard in the legend of King Arthur and the Knights of the Round Table. According to legend, he was the greatest, most powerful wizard in all the land, and so, of course, are you and your team in this exercise. Merlin's magical powers could be channeled in four ways, to enlarge, to shrink, to make vanish and to reverse. But you can always dream up your own variations – for example, the world's cheapest, the world's most expensive, the world's smallest and the world's craziest. When applying the Merlin method, it's very important to think about the types of idea you are interested in generating and how the central issue should be formulated.

Step by step

1. Introduce the central issue.

2. Now demonstrate your first magic trick, preferably with a clear example.

For example: "What would happen if we took our existing products and made them incredibly small? Well, you could put them in your handbag, you could stick them in your computer, you could swallow them..."

3. After about 10 minutes, you should conjure up a new perspective and then continue like this until all perspectives have been covered.

4. Wind up the exercise by allowing team members to write down their ideas while enriching them with information and substance.

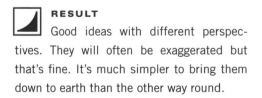

SITUATION
The method is very suitable for advancing and refining existing products. It might be a little tricky to start with this method from scratch.

RESULT
Good ideas with different perspectives. They will often be exaggerated but that's fine. It's much simpler to bring them down to earth than the other way round.

PARTICIPANTS
5–12.

ENVIRONMENT
Anywhere.

TIME FRAME
Allow about 6–10 minutes per perspective, a short introduction and 5–10 minutes to write down ideas = 30–50 minutes in all.

MATERIALS
Post-its, pens or large sheets of paper to illustrate your ideas while you're generating.

YOUR CREATIVE IDOL

Your Creative Idol involves mentally dressing up as a few incredibly creative people that you admire. How would Elvis have solved the challenge that's facing us, or what "glasses" would Leonardo da Vinci have worn to find new ideas around the focus area? You can use a famous person that you've chosen yourself, or some of your team's creative idols. Like What If?, this method has a bit of a role-play feel to it, though with Your Creative Idol, team members are more active in choosing the people that they can identify with.

Step by step

1. Introduce the method (not the central issue) and then suggest a few people whose creativity should be described. You can mix stereotypes and famous people with powerful public images. Impressions will be enhanced if you use photos of the various personalities. You'll get lots of laughs if these images don't only include portraits of famous creators but also pictures of people who look rather boring.

2. Then either split your team into small groups of two to four people or let them work individually. Give them 10 minutes to describe how their "idol" would be able to solve the issue and how their creativity works.

3. Introduce the central issue and let your team ask any questions they have.

4. Ask them to take on the role of their creative idols and solve the problem the way their idols would. Make sure that lots of solutions are generated.

5. To encourage a more dynamic mood, it can be good to rearrange the groups and ask them to work with someone new so that they can test a new role.

6. A final presentation during which both personalities and ideas are demonstrated is always a fun way to wind up this method.

SITUATION

When team members have worked with an issue for a long time and need some distance from their usual approach. Or just an original idea generation method.

RESULT

Relatively tangible ideas and concepts from a range of perspectives.

PARTICIPANTS

6–30.

TIME FRAME

This method can last anything from 30–40 minutes up to several hours. Just make sure it stays interesting and that the groups are constantly thrashing out ideas.

When energy levels start to drop, you should rearrange the groups or wind up the exercise.

ENVIRONMENT

A room for each group to stop them from eavesdropping on each other.

MATERIALS

Paper and pens. It's also an advantage if there are nice, large color images of the people being described, as back up if nothing else.

THE RELAY BATON

The Relay Baton is a simple method that involves continuously building on each other's ideas. Team members pass ideas amongst themselves and challenge each other to develop them by adding new aspects and angles before passing them on. The Relay Baton is a simple technique to apply and quite useful, but – like brainstorming – not particularly original.

Step by step

1. Introduce the central issue.

2. Ask your team to generate ideas on A4 paper either individually or in small groups. Make sure that several ideas are generated and not just one idea per group.

3. Pass all the ideas on to the next person or group and ask them to add a little substance to the ideas. This should be done at a fast pace! After one completed lap, give the original idea originator the opportunity to prepare his or her idea and present it to the group as a conclusion.

SITUATION
When the central issue is relatively uncomplicated. When there's a need for fast, amusing solutions.

RESULT
A small number of reasonably tangible ideas in a short time.

PARTICIPANTS
4–50.

TIME FRAME
5–25 minutes.

ENVIRONMENT
Good possibilities to pass paper to each other. Generally speaking, U-shaped seating is not a good idea during idea generation, but in this case it works reasonably well.

MATERIAL
A4 paper and pens.

ZOOMING OUT

Most of us have dreamed of one day being given the opportunity to travel in space and observe our insignificant, colorful planet Earth from a distance. The Zooming Out method is
most suitable in product development using some kind of product or service as its basis. The
method has similarities with Random Word Association but is more controlled and precise.
In brief, you zoom out from your starting point to achieve a broader view, still keeping your
own product or service in focus.

Step by step

1. Introduce the product or service briefly and concisely.

2. Start by drawing up a list together with your team of all the product or service
features.

3. Make a new list of products that share similar features.

*For example: if one of the features of the product is that it's transparent, make
a list with a lot of other transparent things.*

Continue this process with five to eight features. The result will be a list of feature
categories and a mass of similar products.

4. Introduce the product development needs. What does the product require for its
development? Is it a smart new detail? Does it need tailoring to a new target group
or market?

5. Which of the listed features are essential and how can the products on your list
contribute to the development of your product? Look carefully at the listed products
and their specific solutions and generate ideas around them. Drive the team on so
that lots of ideas are generated.

6. Give the team a moment at the end to write down and enrich their ideas.

SITUATION

When there's something that you want to improve or develop with the help of products from other industries or markets.

RESULT

A mass of product examples from other industries and a number of ideas based on these. A vertical examination of similar products in other fields.

PARTICIPANTS

5–12.

TIME FRAME

About 45 minutes in all.

ENVIRONMENT

Access to a whiteboard or other suitable wall space.

MATERIALS

Paper and pens.

THE LOTUS BLOSSOM

Designing a process that grows from a seed and develops into a flower in full bloom is an apt description of an idea process. The flower has become a symbol for this logical method. The Lotus Blossom is a simple, effective exercise for the rapid creation of solution suggestions that are tangible, yet inspired by other ideas to break new ground.

Step by step

1. Prepare a grid on a large chart like the one in the diagram.

2. Define your need. Formulate it as a question and write it in the center of the diagram.

3. Generate solutions to the need.

4. Draw up a list of the eight best solutions and place them around the central issue.

5. Reflect on the diagram and see if other new ideas can be generated. If some can, write them down and consider where to place them in the diagram. There should only be eight ideas around the central theme.

6. Move the eight ideas outwards to the next square. Write a large How? next to the ideas to help you create more tangible solutions.

7. Generate ideas based on these eight original ideas and create eight new ideas, or more tangible sub-solutions, per original idea. Write them in the grid around their respective original ideas until the lotus blossom is fully developed and complete.

SITUATION
When you need the reassurance of logic and a consistent development of ideas in different directions.

RESULT
Up to 64 tangible ideas with a nice graphic roadmap showing how you achieved your results.

PARTICIPANTS
5–12.

TIME FRAME
About 45 minutes.

ENVIRONMENT
Room to work on a large wall with a whiteboard or a large chart.

MATERIALS
Post-its, pens and very large, prepared lotus blossom templates.

TRENDSTORMING

Fantasizing about the future is often an inspiration engine for visionary thinking. Remember what an impact and self-realizing effect many trends, scenarios and science fiction stories have had on the present day and the future. The Trendstorming method requires you to identify a number of current trends that are relevant to the focus area. Search for general lifestyle trends from authors such as Faith Popcorn or Richard Watson, future studies from various institutes, or analysis carried out by your own organization concerning, for example, new customer behaviors or new technological features.

Step by step

1. Identify, prepare or process a number of trends that might impact on the product being developed or the actual focus area. Ensure that you introduce your trends in a smart inspiring way, either by talking about them or by making information sheets.

2. Write up the focus area on a board so that your team can see it.

3. Split them into groups of three to four.

4. Introduce a current trend in as visual and fascinating a way as you can. Be careful to use tangible examples. Either introduce one trend at a time to all the groups, or introduce different trends to different groups.

5. Then let the groups work on the focus area with the trend as inspiration and guide. You can set a target that each group should generate a minimum number of ideas. Be careful to remind your team that every idea should be written down and that respective idea originators are responsible for documenting their own ideas. You can choose to coach the groups intensively – which will create more energy and spontaneous ideas – or let the groups work more independently, which often results in a more rational mood.

6. Gather up the ideas from each group.

SITUATION

When you know that you have a focus area that needs developing in a certain direction, but also an inspiring and visionary method that produces tangible results.

RESULT

Tangible results within a particular area, and with ideas that are very much in line with current trends.

PARTICIPANTS

3–30.

TIME FRAME

Depending on the number of trends, the exercise can last from 30–60 minutes.

ENVIRONMENT

Several smaller meeting rooms.

MATERIALS

Presentation materials for your trends (images, prototypes, film clips, lifestyle objects, short presentation of each trend, etc.).

UPSIDE DOWN

Have you ever stood on your head and looked at the world from a new angle? It would probably surprise you. Upside Down is an effective method for advancing products and services. As its name implies, it involves turning things upside down. For example, you can invert the features and functions of a product or service, or even the physical product itself.

Step by step

1. Prepare a number of feasible Upside Down scenarios – which features are useful and how to turn them on their heads.

2. Start by describing the focus area or the product.

3. Then together with your team, describe the desirable features of the product or service.

4. Then try to invert the functions of the product. Guide your team by giving examples.

For example: The subject is the development of a new rubber boot.

Waterproof – How would a non-waterproof rubber boot look? How can we apply this function?

Comfort – Can we make the boots more uncomfortable? What could we gain from this?

Internal moisture – Can we get the moisture on the outside? What material should we use?

5. Make sure that all ideas, no matter how crazy they are, are written down!

SITUATION
When there's already a subject to apply as a basis, for example a product, or you're bogged down in a stale approach.

RESULT
Often provides new and exciting angles and generates useful ideas.

PARTICIPANTS
5–12.

TIME FRAME
About 45 minutes.

ENVIRONMENT
Make sure that there's a whiteboard available or a flipchart to list the features on.

MATERIALS
Paper and pens.

TRIZ

TRIZ or TIPS (acronym for "Theory of Inventive Problem Solving"), is a method created by the Russian Genrich Altshuller. There is a lot written about TRIZ, as the methodology is comprehensive, but in brief it is a system for technical problem solving. Altshuller analyzed thousands of patents and technological innovations and found that there were clear commonalities that could be summed up in 40 so-called innovative principles. The TRIZ methodology in its simplest form may be used to develop technical products. When you have stated your problem in the need phase, you can use the 40 principles as a sounding board to see if they can create new solutions. The following describes how to use TRIZ in an easy way, which can be tough enough for those who do not have an engineering background. If you want to use TRIZ in a more sophisticated form of computations and calculations, you need to hire specialists in the Triz methodology.

Some examples of innovative principles:

Some more, some less. What happens if you add or subtract substances in the product?

Asymmetry. What happens if you change symmetric parts of the product into asymmetric?

Flexible membranes and thin layers. What happens if you replace traditional or rigid parts with flexible parts?

Thermal expansion. What happens if you use the effect of expansion of different materials at different temperatures?

Hollow materials. What happens if you use hollow or porous elements (shells or add ons)?

Horizontal motion. What happens if we limit or alter the horizontal movement?

Powerful oxides. What happens if you replace the air with an oxygen-rich, or similar, solution?

Separation. What happens if you separate the various elements or characteristics of the elements?

Step by step

1. Before the assignment, you as an Idea Agent should discuss with the need owner which innovative principles you believe are relevant to work with on the specific need. Print the principles you will use in a clear and visible format and describe them in a simple way as shown above.

2. Tell participants about the technical challenge or problem that you are facing.

3. Make a number of stations on the tables. On each station you put out a printed sheet of one innovative principle. That is, one principle per station. Place Post-its on each station.

4. Let the participants go to a station in pairs. Ask them to think briefly about the innovative principle before them and then, with the principle as a basis, find new solutions to the issue. They then to write down ideas on Post-it notes while talking and before they leave every station.

5. Tell the participants to shift to the next station after about five minutes and go in their pairs to a new station.

6. Continue until all pairs have been at all stations or until you feel that the groups need a pause (usually after about 30 minutes).

 SITUATION
Good for technical needs. TRIZ does not always create clear answers but it is a way of using a systematic scientific base in order to seek inspiration to new solutions and ideas.

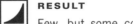 **RESULT**
Few, but some concrete ideas and suggestions.

PARTICIPANTS
2–15 technicians. TRIZ is difficult to run without participants with an engineering background.

TIME FRAME
30–60 minutes.

ENVIRONMENT
Preferably a room where you can build a lot of small stations for two people to work at.

MATERIALS
Printouts with the innovative principles. Post-it notes and pens.

THREE DOTS

Three Dots is a simple, informal selection method for ranking ideas that involves team members voting for their favorite ideas in relation to three criteria. The method is practical when you need a quick evaluation of ideas based on the gut feelings of your team. It can also be applied as an initial screening method.

Step by step

1. Put all your ideas – as Post-its or on sheets of paper – up on a wall or spread them on the floor so that your team can get a good overview. Ask them to walk round and take in the ideas as they might in an art exhibition.

2. Give each team member a predetermined number of self-adhesive dots in three different colors, with each color representing a predefined criterion.

For example:

Blue = a rapidly realizable idea

Red = an innovative idea

Green = good market potential

3. Let your team walk round grading which ideas they think are the best in each category.

4. Preferably wind up the exercise with a quick review of the best ideas to see if there's anything new you can add to them.

SITUATION

When you'd rather screen ideas in an uncomplicated and relaxing way. When the central issue is not crucial to the future of the company, but the project owner requires assistance in the process follow-up. The method works well as a conclusion to the day's idea generation and will seem even more like an art exhibition if you have a glass of wine in your hand.

RESULT

All the ideas are assessed by the team on the basis of three criteria. The project owner gets a quick assessment and an internal prioritization of ideas that is relatively clear. This method is also easy to use with a legend to present your ideas later on in the process.

PARTICIPANTS

3–20.

TIME FRAME

2 minutes of introduction + 10–30 minutes depending on the number of ideas.

ENVIRONMENT

A wall that can function as an idea gallery.

MATERIALS

Small self-adhesive colored stickers (why not gold stars?) or colored pens.

TOP 10

The inspiration for Top 10 comes from the selection of the top 10 songs in Billboard's Hot 100 singles chart, in which pop songs are treated as commodities and continually assessed by the market. Top 10 is a simple evaluation method that can be applied to select a specific number of ideas from the pool. It could very well be called Top 20 – you can choose the number of ideas that are relevant to your process. You can also spice up the method using terms such as "bubbling under" and "bullet of the week".

Step by step

1. Put all the ideas – as Post-its or on sheets of papers – up on a wall so that they can be viewed. Let everybody walk round and take in the ideas.

2. Make a list from 1 to 10 on the wall.

3. Each team member should now take an idea from the idea pool (i.e. the mass of ideas on the wall), rank it and justify its spot on the list. If a discussion starts or an idea is developed during this process, it's important to document any additional thoughts.

4. Continue the discussion while moving the ideas up and down on the list until the team has a list they are satisfied with.

5. The end result will be a Top 10 "chart" of selected and hopefully reasonably developed ideas that have been internally ranked.

SITUATION
The method is most suited to ideas that are easy to evaluate, but less to complex central issues or ideas that are hard to assess.

RESULT

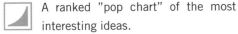
A ranked "pop chart" of the most interesting ideas.

PARTICIPANTS
3–8 per group.

TIME FRAME
2 minutes of introduction + 5 minutes of reconnaissance + 15–30 minutes (depending on the number of ideas) in groups.

ENVIRONMENT

Empty wall space to put up your ideas.

MATERIALS

Post-its, a large smooth wall to put your ideas on, or a whiteboard on which to write your list. Scotch tape to stick up ideas written on regular sheets of paper.

_____

_____

_____

_____

_____

_____

_____

_____

_____

THE FOUR-FIELD MATRIX

The Four-Field Matrix is a method used to rank the usability of ideas and is probably the most common of all the screening and evaluation methods. Nine- or 16-field matrices are also frequently used to achieve a greater differentiation and to increase opportunities to apply more values on the various axes. It is important to emphasize that this should not only be seen as a placement exercise, but one that enables you to transfer (and enrich) as many ideas as possible into the optimum fields.

Step by step

1. Carefully think through, or ask the project owner, which parameters to apply for evaluating your ideas using the Four-Field Matrix. Typical parameters are: risk, realizability, cost, or level of innovation (see p 77). Select two parameters and allocate two assessment levels to them – for example, high risk and low risk, or high innovation level and low.

2. Draw the matrix on a whiteboard or a large sheet of paper with four fields of equal size in a large square (see fig. A). Write the parameters next to their respective fields.

3. Introduce the exercise and explain why you have chosen these particular parameters. What is your purpose for evaluating the ideas using these parameters?

4. Let team members place ideas in the matrix fields that they consider most appropriate. The best effect is achieved if this is done in groups by individual idea, but if time is short it can be done individually.

5. When all the ideas have been placed in the matrix, try to ensure that they've been accurately assessed. Does the team agree on the placement of ideas – in other words, the assessment of ideas – in the matrix? If results are acceptable, then it's time to advance the best ideas to the next stage of the process, but you usually need to apply two or three matrices to pick out the best ideas based on your criteria.

Our example is based on the focus area: "How could a 24-hour service for a county authority be organized?"

The authority's strategy was to rapidly realize a concept for 24-hour servicing of its citizens. The most important parameters for idea assessment were therefore: Will the service be accessible 24 hours a day? How long will it take for the authority to create it?

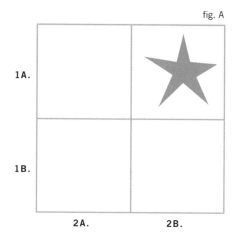

fig. A

1A.

1B.

2A.   2B.

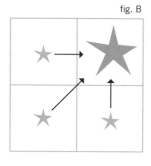

fig. B

1A. CAN BE DONE WITHIN 6 MONTHS.

1B. WILL TAKE MORE THAN 6 MONTHS TO DO.

2A. CAN'T BE ACCESSED 24 HOURS A DAY.

2B. CAN BE ACCESSED 24 HOURS A DAY.

6. If results are unsatisfactory – in other words there are too few ideas placed in the optimum field in the matrix – then it's essential to examine the existing ideas and try to enrich and advance them on the basis of the selected criteria to enhance their quality (fig. B). This is done most effectively through discussions in small groups.

SITUATION

The Four-Field Matrix is a technique that is used to pick out the ideas that are most relevant to the focus area and the strategy of your organization. It's a common method for processing and developing a large number of ideas quickly and assessing them accurately.

RESULT

An idea map that provides a good overview and a relevant assessment of ideas and concepts. However, to do your ideas justice, you should apply several matrices with contrasting axes. The method also develops ideas from a "Post-it level" to a more tangible level.

PARTICIPANTS

1–8.

TIME FRAME

5 minutes of introduction + a minimum 25 minutes in groups. Can last several hours depending on the number of ideas and how much focus is put on idea development.

ENVIRONMENT

A large wall space with a large sheet of paper or a whiteboard on which to draw the matrix.

MATERIALS

The Four-Field Matrix requires the use of Post-its or an abundant supply of scotch tape. Large numbers of pens and large sheets of paper should be made available to allow room for idea development.

THE SPIDER WEB DIAGRAM

The Spider Web Diagram is a method that provides a much clearer picture of the value of different ideas than matrices with columns of figures because it is both visual and can apply more than two parameters at a time. Graphic matrices are often effective when you want to compare different ideas or present them. Try to provide a composite picture of the end product during your presentation by writing ideas in different colors. The finished web creates a good overview of the result.

Step by step

1. Define your assessment criteria (usually between four and six is preferable and relatively easy to work with). This can be done in advance, but a discussion is often constructive and can develop a more detailed picture

2. Examine and assess your ideas one by one on the basis of the defined criteria. Mark the assessment results for each idea on the diagram. Preferably give each idea a different color.

3. Discuss the results and determine whether there are any recurring trends. Collect the results and rank your ideas on the basis of the diagram outcome.

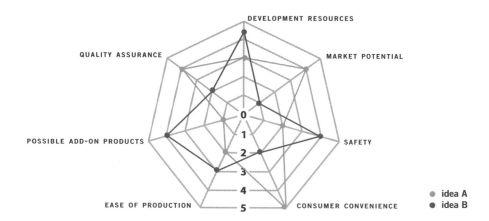

SITUATION

When there is a limited number of relatively tangible ideas to assess and you want a graphic overview that's easy to understand and present.

RESULT

A visual overview that is simple to present and in which the relative strengths and weaknesses of each idea are easily identified. Its visual nature enables the easy identification of trends.

PARTICIPANTS

1–6.

TIME FRAME

From about 15 minutes per idea upwards depending on complexity of ideas and the amount of time needed for accurate assessment.

ENVIRONMENT

A room with a lot of empty wall space to put up your graphic results.

MATERIALS

It simplifies matters if you prepare several diagrams in advance on large sheets of paper so that it's easy to fill in results. Different colored pens are needed as well.

KESSELRING

In product development contexts, a wide variety of screening tools have been used through the years. One of the most well known is Kesselring from 1951. It consists of a table with a number of defined selection criteria on one axis and ideas or concepts on the other. Kesselring has inspired a number of followers, among others Pugh's concept selection matrix from 1990 and Ulrich and Eppinger's model for concept selection from 2000.

To achieve success with such an advanced idea evaluation matrix requires investing time in discussing and defining the criteria that will form the basis of your proposed selection. If you choose the more advanced model (see model B on the next page), these criteria must also be weighted before the criteria selection process is complete.

Step by step

1. Begin by describing the ideas so that your team is fully aware of what they'll be assessing.

2. Define your criteria. Usually there'll be some criteria that you've applied previously in four- or nine-field matrices and these should obviously be applied here as well. This method often produces new, more detailed criteria that you, in collaboration with the project owner, have kept in reserve from start up or during the idea process.

3. If you choose model B, weight your criteria internally.

4. Assess your ideas in relation to the parameters and document the results in the table. A figure-based points system is most common, but you can also use pluses and minuses, or different color-coding systems.

5. Try to summarize or form an impression of the best ideas.

6. If the central issue or need is very important, or if the ideas are not of sufficiently high quality, try to develop their weak points.

model A

IDEAS AND CONCEPTS			
	Idea A Battery	**Idea B Combustion**	**Idea C Solar cell**
Selection criteria	Points	Points	Points
Development resources	3		
Market potential	4		
Safety	2		
Consumer convenience	5		
Ease of production	2		
Possible add-on products	1		
Quality assurance	4		

Total points	21		
Ranking	1		
Continue with idea? Y/N			

model B

IDEAS AND CONCEPTS							
		Idea A Battery		**Idea B Combustion**		**Idea C Solar cell**	
Selection criteria	Weight-ing	Points	Weighted value	Points	Weighted value	Points	Weighted value
Development resources	15%	3	0.45				
Market potential	25%	4	1				
Safety	10%	2	0.2				
Consumer convenience	15%	5	0.75				
Ease of production	10%	2	0.2				
Possible add-on products	10%	1	0.1				
Quality assurance	15%	4	0.6				

Total points	3.3			
Ranking	1			
Continue with idea? Y/N				

SITUATION

When you want to make detailed comparisons between different ideas and concepts. The method requires that the idea process has progressed a long way and that there is documented information and thoughts on various aspects of the ideas. The ideas should have reached the stage in which they exist on a sheet of A4 or possibly a sketch.

Applying this method as a screening technique with undeveloped "Post-it ideas" is not recommended. It is best suited as a concluding comparison of 3–10 interesting ideas.

RESULT

An objective, detailed and quantified comparison of ideas, whose purpose is to provide a clear impression of how different ideas relate to each other and, if desired, how they relate to a reference idea. It also provides an effective documented summary that can be applied later in the process during presentations or as a decision basis.

PARTICIPANTS

3–6, preferably experts or people with a good working knowledge of the subject.

TIME FRAME

From one hour upwards depending on how well prepared you are and the number of ideas and criteria that need to be assessed. You will often need to invest a half to a whole working day split into sessions to complete the process. Don't be wasteful!

ENVIRONMENT

A room with a work table and chairs, and a large wall space for the matrix itself. A whiteboard or wall space for a large sheet of paper that makes it easy to work and for everyone to see.

MATERIALS

Pens, paper and prepared templates for execution and documentation.

CIRCLE OF PORTRAITS

Circle of Portraits is an excellent method of enabling your team members to notice each other, while raising the temperature in the team and with the help of colors and shapes releasing a little creative energy at the same time. The method contributes to creating a pleasant and light-hearted mood. It also results in a large number of portraits that bear a striking resemblance to role models and which underpin the continued process. Drawings should be done spontaneously and impulsively as a warm up to help the creative flow. One way of increasing its potential is to play relatively uptempo music during the exercise.

Step by step

1. Half of your team should sit in a circle on chairs or on the floor facing outwards. The other half should sit opposite them facing inwards. There should now be two circles of team members sitting in pairs and facing each other.

2. Make sure that every pair has a sheet of paper and that the team members in the outer circle each have a colored pen.

3. On a given signal the team members in the outer circle start drawing a portrait of the person sitting opposite them on their sheet of paper for about 10–30 seconds. You then ask the team to change places, rotating the outer circle one place to the right. The outer team members leave their drawings behind them, but take their pens, and start drawing the new person sitting opposite them on the portrait that the previous team member has started. The exercise continues until they have all gone full circle. The important thing is not to let the process slow down, but to keep drawing all the time.

4. When the team has gone full circle, team members in the outer circle should then reach forward and introduce themselves to their models. They should then write the model's name on the portrait and hang it on the wall with scotch tape.

5. The outer and inner circles then change places and the whole process starts again from step 2.

SITUATION
A good method to use at the start of the process to develop a pleasant and relaxed atmosphere quickly. Creating a sympathetic and cheerful mood at the start of idea generation is one of the keys to enabling team members to relax and pluck up the courage to produce crazy, provocative ideas.

 RESULT
The exercise will get lots of laughs and help the team to bond. The physical results will be lots of absurd portraits that can decorate walls and break the ice.

 PARTICIPANTS
8–20.

 TIME FRAME
2 x 5–10 minutes.

ENVIRONMENT
A large room with enough space for a large circle of people with tables and chairs or down on the floor. Wall space to put up one portrait per team member.

MATERIALS
A3 or A4 paper. Colored pens. Music. Scotch tape.

HERE'S MY KEYRING

This is a simple and slightly unusual introduction technique to get your team warmed up and revealing slightly more personal details than they usually might. Starting a creative session with the same old round of introductions can not only bring down your team's energy levels, it will also focus on the "experts" in the team when they tell each other their job titles. Here's My Keyring usually results in team members starting the session with a collective laugh.

Step by step

1. Ask team members to take out their keyrings with all their keys on it.

2. Start by introducing your own keyring (for example the three most important keys). Say what the keys can open and why they are important to you on a personal level.

3. Ask the other team members to do the same until everybody has completed the drill. Hopefully, the group will now have gotten a more spontaneous introduction to each other than would normally be the case.

 SITUATION
When there are new members in the team or when they're all meeting for the first time.

 RESULT
Personal and often amusing introductions that create a nice mood and a fertile basis for the process in a relatively short period of time.

PARTICIPANTS
Around 10 at most, otherwise even this introduction can get a bit boring.

 TIME FRAME
10–15 minutes.

ENVIRONMENT
Can take place anywhere.

MATERIALS
The team members' key rings. If some of the team members don't have their key rings with them, they should introduce them from memory. It's not the physical keys that are important here, but the metaphors and symbolism.

"ALLOW ME TO INTRODUCE..."

For those of you who feel you've endured far too many uninspiring team introductions, this exercise might do the trick. It enables team members to form a quick impression of each other, which contributes to a high energy level in the room. Each of the team members gets their moment in the limelight and the chance to give the rest of the team a good ole proverbial pat on the back.

Step by step

1. Each team member takes turns to stand in front of the team and introduce one of the other team members in the same way as an announcer would at the Super Bowl. The atmosphere generated should be like that at a large sports event.

For example: "Ladies and gentlemen! I have great pleasure in introducing you to a fantastic person who is incredibly talented at.....and who, moreover, is a virtuoso at.....and to top it all is one of the nicest people I've ever met. A big round of applause for.....!"

2. The Idea Agent kicks off the exercise. The team member who's been introduced then stands up in front of the team and takes their acclaim in the form of applause, stamping and loud cheering.

3. When the applause has died down, the presented team member then continues the exercise by introducing the next team member. If you don't know about much them, you should find out a little in advance or just make it up as you go along.

SITUATION
During the introduction of a creative team and its members to each other. This is an exercise that works well in most contexts even if the individuals in the team don't know each other that well or they're a bit shy.

RESULT
Laughter, loud cheering, applause and a positive mood.

PARTICIPANTS
5–25.

TIME FRAME

About 1 minute per person.

MATERIALS

No specific materials are required.

ENVIRONMENT

Preferably a large open room.

BODYCOUNT

The name of this exercise sounds rather more gruesome than it actually is because the aim of Bodycount is just to count using your entire body. It's a very practical exercise to have up your sleeve when you feel that your team needs a short intensive kick. Bodycount doesn't take long but it generates a good deal of positive energy. Make sure that you set a good example yourself and that the volume is loud – don't chicken out!

Step by step

1. Make sure that all your team are standing up and have a little elbowroom.

2. Start counting out loud from one to eight as you wave your right arm over your head an equal number of times. Then kick out with your right leg eight times, followed by your left leg eight times.

3. Repeat the process but this time only to four. When you've gone through your whole body, count to two and finally to one.

4. Wind up the exercise by first touching your toes then quickly stretching your arms above your head at as you shout out a very loud "wwwhhaaayyttogggo".

 SITUATION
This exercise is very effective when your team is bogged down and needs a quick, physical boost. Try to use it after a long period of physical inactivity.

 RESULT
The exercise revitalizes the energy and mood in the team and releases mental and physical tension.

PARTICIPANTS
3–300.

 TIME FRAME
3 minutes.

ENVIRONMENT
Anywhere.

 MATERIALS
No specific materials are required.

HOME ALONE

Home Alone is a very fun way to boost team energy levels and release a little tension at the same time. It's inspired by the scene in the film *Home Alone* in which young Macaulay Culkin takes aftershave in the palms of both hands and claps it against his cheeks. He then opens his mouth as wide as it will go and shouts at the top of his lungs. Home Alone works most effectively in large teams and with relatively uninhibited people. It is important that the Idea Agent demonstrates the exercise enthusiastically so that team members can see that it's OK to let loose.

Step by step

1. Your team should stand in a circle with their heads bowed.

2. When you give the signal, they lift up their heads and look at the team member of their choice. If two team members make eye contact, they should clap their hands to their cheeks, raise their eyebrows, open their mouths as wide as possible and shout as loud as they can.

3. Those that have shouted then leave the circle and the rest bow their heads again.

4. The exercise then starts again from step 2 until everybody has shouted at least once.

 SITUATION
When you need to generate more energy in the group or when you want to try an off-the-wall icebreaker.

 RESULT
Laughter and a higher pulse.

 PARTICIPANTS
5–30.

 TIME FRAME
2–10 minutes.

 ENVIRONMENT
Preferably an open area.

 MATERIALS
No specific materials are needed.

STORM WARNING

To really get your team going often requires focusing on the most important bodily function of all – your breathing. Storm Warning is an exercise that guarantees a large intake of oxygen in the body, and it also clears the mind very effectively. The competitive element will help your team to think strategically.

Step by step

1. Split your team into two groups with a maximum of four in each group.

2. Then draw a line or a piece of string between them – this boundary should not be overstepped.

3. One or several eiderdown feathers should then be released between the two groups at eye level with both groups then blowing to try and prevent the feather(s) from landing on their side of the line.

4. After each round both teams have 30 seconds to think up a strategy that will win the next round.

5. The overall winners are the first team to win three rounds.

 **SITUATION**
When physical energy is needed in the team. There is also a strategic dimension to this exercise that can be brought up and discussed. How did the teams go about solving the exercise? What seemed to be the best winning strategy?

RESULT
Stimulates the team members and their bodies, both physically and mentally. In addition, it provides an insight into the roles of different personalities in the team.

PARTICIPANTS
Anyone can take part – except people with breathing difficulties. 4–8 people.

 **TIME FRAME**
2–10 minutes.

ENVIRONMENT
Preferably an open indoor area.

 **MATERIALS**
A rope or a piece of string. 1–5 feathers that float and won't fall too quickly.

THE ROLLER COASTER

Taking a really great roller coaster ride is one of the most exhilarating and energizing experiences we can have. This exercise recreates the adrenalin and power of the roller coaster experience. Like Bodycount, it's an effective method to keep up your sleeve for when energy levels are low and you need a quick way to revitalize your team and their mental processes. Make sure you practice the sound and movement of a roller coaster before you get started – it's important to have an enthusiastic driver!

Step by step

1. The team sits in pairs shoulder to shoulder in a long line like a roller coaster. As the Idea Agent, you should sit at the front and steer.

2. Your task is to simulate the roller coaster ride. By making roller coaster sounds, yelling loudly on descents, leaning backwards on ascents, swaying sharply at curves and holding on for dear life, you steer the roller coaster around its circuit. Your team should just enjoy the ride and imitate you.

3. The ride ends when the roller coaster arrives back at its starting point.

 SITUATION
When you need to generate energy, either at the start of the idea process or for periods of low energy during the process.

 RESULT
An energy boost.

 PARTICIPANTS
5–25.

 TIME FRAME
5–10 minutes.

ENVIRONMENT
Chairs set out like a roller coaster.

 MATERIALS
Possibly a megaphone.

CONCENTRATION TO 20

Enabling your team to achieve complete focus during an idea process is vital. Concentration to 20 is a method that helps improve focus and concentration within the team. Your team members will be reminded that they have been gathered together for a single purpose. Recognize what is happening in the team and prevent them from solving a problem using patterns – focus and intuition are the elements that should be highlighted.

Step by step

1. Ask the team to form a circle either sitting or standing and close their eyes.

2. They should now count to 20. One of the team members starts the process by counting "one", then one of the others counts "two" and so on. However, there are a few rules:

- Only one person can count each number – in other words, two or more should not say the same number at the same time.
- They must not create patterns, for example circular counting – the counting should take place at random.
- Nor are they allowed to make signals to each other.
- Individual team members can count several numbers, but not two in a row.

If any of the team breaks these rules, they have to start over.

3. When the group has managed to count to 20 without any lapses, they'll have finished the exercise!

SITUATION
An effective exercise when your team lacks concentration and focus.

RESULT
A more focused team.

PARTICIPANTS
5–15.

ENVIRONMENT
An open space and possibly a circle of chairs.

TIME FRAME
It depends on how quickly the team completes the task – it can vary from 5 to 30 minutes.

MATERIALS
No specific materials are required.

THE DUNCE HAT

The Dunce Hat is an energy generation method that also provides food for thought with regard to making conscious choices. As Idea Agent, you should keep a firm control on the exercise and act as referee. Observe and document any interesting details and bring them up for discussion during the subsequent review.

Step by step

1. Each team member makes their own dunce hat, which should fit properly on their head, using a piece of A3 paper and scotch tape.

2. Ask them to choose a partner and then give them these instructions: The exercise is a contest to see who can steal their partner's hat or to knock it to the ground first – first to five points wins. Rules:
- they can only use their hands offensively.
- they can move their bodies both offensively and defensively.
- stolen hat = plus 1 point.
- dropped hat = minus 1 point.

3. Assemble the group at the end of the exercise and discuss the following questions: What strategies were applied during the contest? Which proved to be more effective, offense or defense? Did team members make any conscious choices? The offensive strategy usually wins in the long run – the more committed you are, the more you'll succeed!

SITUATION
Before initiating an idea generation method, your team needs to appreciate that an active input will bring rewards. A method that allows a "physical timeout" and reflection.

RESULT
A cheerful, relaxed team that understands in concrete terms that it pays to take your best shot.

PARTICIPANTS
2–16.

TIME FRAME
5 minutes of introduction + 5 minutes of competition + 5–10 minutes of reflection.

ENVIRONMENT
A few square yards for each pair.

MATERIALS
A3 paper and scotch tape.

THE LAST SUPPER?

The Last Supper? is an exercise in generating ideas out of different objects and then trying to develop each other's creations. It is our own variation on the popular daytime cookery show "Ready Steady Cook". The advantage of this exercise is that the result – as well as the training itself – is very tangible to the team. Poor ideas and poor idea development lead to poor food and vice versa! It also involves daring to allow your pet ideas to be assessed and developed by others. The Idea Agent should also try to activate team members who can't cook.

Step by step

1. Purchase a few cookery ingredients without imagining the end result – the meal. Unwrap them in a kitchen.

2. Ask two team members to go into the kitchen and start cooking. They can use the whole kitchen space and all its utensils for cooking, but only the ingredients that have been laid out.

3. After 10 minutes, the first pair should leave the kitchen and a new pair should go in without the two pairs communicating with each other. The new pair should now try to develop the dish that the first pair has started.

4. The process continues until all the pairs have had their turn in the kitchen or until the meal is ready.

SITUATION
When you need a relaxed, unconventional team exercise that is also a practical demonstration of your team's creative ability. It is most suitable when you have plenty of time during the idea process, and during residential courses or workshops.

RESULT
An unusual, creative team-building exercise that produces an exciting meal. Sometimes it might be a good idea to have a reserve meal prepared in case the exercise is a culinary disaster.

PARTICIPANTS
6–12.

TIME FRAME
30–60 minutes.

ENVIRONMENT
A kitchen.

MATERIALS
Make sure you don't purchase ingredients for a particular recipe and try to choose ones that will put team creativity to the test. Also try to have a few alternative ingredients in reserve in case the end result is inedible.

TEN THOUSAND RED ROSES

Understanding the human need for praise and professional affirmation is fundamental in all idea processes. Working creatively involves breaking the mold and daring to venture outside the boundaries of conventional thought. And facilitating this courage requires continuously appreciating the ideas that are generated even if only one in 200 is brilliant. This is why it's important that you wind up an idea session in such a way that your team members will want to keep on spouting ideas. Ten Thousand Red Roses is a perfect means to this end.

Step by step

1. Half of your team forms an inner circle and the other half an outer one. They should stand facing each other in pairs and within arm's length of each other.

2. The team members in the outer circle begin by praising their partners. They can do this by saying how fantastic their partners were during the creative session, what their strengths are, or even how nice their clothes look.

3. After about 15 seconds, the team members in the outer circle take one step to the right and continue the exercise by paying homage to the next team member in much the same way. The exercise continues until everyone has gone full circle. Playing upbeat music at the same time is a good way to raise the mood. Also bear in mind that praising someone can be difficult and slightly sensitive, so treat your team gently.

4. The inner and outer circles then swap places and start again from step 2.

5. Try to wind up the exercise with a little champagne to create even more of a party atmosphere.

SITUATION
At the end of the day or a creative session, or when team members need a bit of a confidence boost.

RESULT
A relaxed atmosphere and a positive ending to the work day.

PARTICIPANTS
6–50.

ENVIRONMENT
A large open area.

TIME FRAME
5–15 minutes.

MATERIALS
A CD player and music.

APPENDIX

A. SOURCES OF INSPIRATION

B. GLOSSARY

C. INDEX – QUICK GUIDE TO IDEA GENERATION METHODS

D. MODEL SCHEDULE FOR A DAY-LONG WORKSHOP

E. LAZY GUIDE TO CREATIVE PROCESS DESIGN

A. SOURCES OF INSPIRATION

A huge assortment of books on the subject of creativity and innovation, written from a variety of perspectives, has been published over the years. It would obviously be an impossible task to list all of these publications (of varying quality) among our sources of inspiration. However, as a compass in the jungle of books on the market we recommend for your further attention the following brief selection of knowledge sources from the world of creativity and innovation.

BOOKS

Amidon, Debra M. (1997): *Innovation Strategy for the Knowledge Economy – The Ken Awakening.*
> – a book of reflections on the internal relationships between knowledge and innovation and how to manage these parameters in large organizations.

Ayan, Jordan (1997): *Aha! – 10 Ways to Free Your Creative Spirit and Find Your Great Ideas.*
> – a first-rate book on how to understand and utilize your creative powers on a personal level.

Christensen, Clayton M. (2001): *Harvard Business Review on Innovation.*
> – a book of short articles and summaries of texts from significant innovation thinkers such as Clayton Christensen, Chan Kim and Renée Mauborgne. The book mainly deals with innovation on a theoretical level, not a practical or hands-on oriented one.

Dundon, Elaine (2002): *The Seeds of Innovation.*
> – a book that is extremely precise and well defined in its discussion of the diffuse notion of innovation, and one that every development manager should read for its tangible message and appreciation of how to create an innovation culture.

Edvinsson, Leif (2002): *The Corporate Longitude.*
> – a book by the father of intellectual capital is always of great interest to an idea person. Read it to enhance your understanding of the world of contemporary and future ideas.

Florida, Richard (2002): *The Rise of the Creative Class.*
> – the most comprehensive sociological study of the rise of the new class – creators. Florida suggests that this new class is defined by the fact that it is paid to solve

problems and create solutions for new opportunities. In contrast, the service and industrial sectors earn their living working on projects that other interests have defined. If you have a special interest in Europe, in 2004 Florida published a special research report on the EU area.

Govindarjan, Vijay and Trimble, Chris (2010): *The Other Side of Innovation.*

– Alexander "Mr Business Model Innovation" Osterwalder's book tip to his clients. The book describes in the abstract way, yet with concrete cases, how different types of innovation require different organizational formats.

Higgins, James M. (1994): *101 Creative Problem Solving Techniques. The Handbook of New Ideas for Business.*

– a hotchpotch of reflections, brief case studies and methods that contains quite a few nuggets if you can sift them out.

Kao, John (2002): *Jamming – The Art & Discipline of Business Creativity.*

– a Harvard professor that has become an idea guru and who writes about how we entered the age of creativity. Most interesting for its hard-hitting metaphors and the power of its business intelligence.

Kirton, Michael (2003): *Adaption-Innovation: In the Context of Change and Diversity.*

– the man behind the KAI-test, Dr Michael Kirton, discusses his theory of creative personality styles and how to apply it to improve cooperation in groups, with a focus on working with problem solving and creativity.

May, Rollo (2002): *The Courage to Create.*

– in all probability creating is one of the essential human elements in contemporary society, a society in which more and more people are seeking to realize themselves and their potential. Psychologist Rollo May writes in a captivating way about the joy and angst of the creative process.

Miller, William C. (1999 / 2000): *Flash of Brilliance / The Flash of Brilliance Workbook.*

– a former research consultant at Stanford Research Institute (SRI) who first wrote an excellent philosophical observation of innovation processes in organizations and then clarified his practical thoughts in a workbook.

Sternberg, Robert J. and James C. Kaufman (2010): *The Cambridge Handbook of Creativity.*

– an excellent review of the research world's perspectives on creativity. The book follows and summarizes various research areas such as psychology, history, cognitive science, organization theory and biology to provide a comprehensive picture of the elusive substance, creativity.

Strøm, Ole (1992): *Idéudvikling*.

– a Dane who writes in a practical and tangible way about idea development and its methods. The book is of limited creative value in form and style, but under its dull exterior it can frequently prove quite handy.

Ulwick, Anthony W. (2005): *What Customers Want*.

– a book that is critical of the traditional way to work with market needs, which is to ask customers what they want. It describes instead a method of self-analysis to identify opportunities for improvement.

Williams, Luke (2011): *Disrupt: Think the Unthinkable to Spark Transformation in Your Business*.

– a book that describes the process design of the firm Frog Design when they want to create unexpected and completely new products. Much focus is on how to identify and formulate the need.

WEBSITES

www.innovationmanagement.se

– a Swedish venture with an international outlook that takes innovation seriously. It focuses on both research and practice and blends methodology with straight up interviews.

www.innovationtools.com

– a blog focused on innovation management and creativity techniques. Innovation Tools can sometimes feel a little old school American but it contains much information if you are ready to do some searching.

Words and their meanings can often cause serious communication setbacks between individuals in general and in educational contexts in particular. A glossary that records how the authors of this handbook define its essential terms is therefore appropriate. But please note that these definitions are the authors' interpretations only and not the accepted wisdoms of Encyclopedia Americana and its equivalents.

BRAINSTORMING – an idea generation method created by the American advertising manager Alex Osborn in the 1950s. Brainstorming has become almost synonymous with creative processing but is actually a relatively simple method that involves a team of people sitting round a table, one person being responsible for taking notes while the others "storm" ideas.

CONCEPT – a defined, developed and tangible idea.

CREATIVE TEAM – a group of individuals assembled to generate ideas and concepts.

CREATIVITY – conceiving and creating – from the Greek word "cre" meaning "to create". A primeval human urge whose significance in contemporary society appears to be booming.

ENRICHMENT – developing an idea into a concept.

FOCUS AREA – a defined need usually characterized by a central issue that steers the idea process in the right direction.

IDEA – a thought that can create reality.

IDEA AGENT – the facilitator of a creative process who applies professional methods to generate ideas and develop them into finished concepts.

IDEA DEVELOPMENT – the phase of the creative process that focuses on gathering, organizing and advancing existing ideas. During this phase, new ideas can be synthesized when combining or confronting existing ideas.

IDEA GENERATION – the phase of the creative process that focuses on creating as many ideas as possible.

IDEA MANAGEMENT – a work structure and an approach to producing, managing and developing ideas within organizations.

IDEA ORIGINATOR – the individual who conceived the idea.

IDEA POOL – the mass of ideas that results from the idea generation phase.

IDEA PROCESS – a general idea process comprises five separate phases: need definition, idea generation, idea development and screening, idea enrichment, and result.

INNOVATION MANAGEMENT – a structure and an approach to developing innovation in organizations: from a supervisory strategy for behavior and available resources, to specific goals such as the proportion of annual sales that should constitute new products.

OPEN INNOVATION – the term was coined in 2003 by Henry Chesbrough, a professor at the University of California. The concept can be defined as an organization's development of ideas through collaboration with people outside the organization, such as customers, vendors, partners, research networks or even competitors.

OUT-OF-THE-BOX – out-of-the-box ideas are solutions to an issue that are beyond the commonplace and the trivial. These are solutions that lie outside conventional frameworks – frameworks that we have often created for ourselves in our imagination but which don't exist in reality. Out-of-the-box ideas may often seem crazy or absurd at first glance, but can be laughably simple, obvious or logical in retrospect.

PROCESS – a course of events, an evolution or a progression. The general idea process comprises five separate phases: need definition, idea generation, idea development and screening, idea enrichment, and result.

RAPID PROTOTYPING – a technique to rapidly produce parts for a model or prototype often using some form of computer aided design (CAD). Rapid prototyping is needed to quickly get a real feel for a product to enable feedback from customers in early stages of the development process.

SCREENING – processing and sifting a pool of ideas on the basis of defined relevant criteria.

STAKEHOLDERS – individuals or groups with an interest in the success of an organization or a creative process in delivering intended results.

WORKSHOP – a work process during which a group of people concentrate on a project in a specific location for a specific period of time. A workshop is realized most successfully under the guidance of a facilitator.

The intention of this quick guide is to simplify your search for and selection of idea generation methods. An overview such as this will rarely provide complete classification accuracy, but can be applied as a quick introduction to the method or methods that suit you best. Begin by considering the need starting point, expectations for the result and your own skill as an Idea Agent.

QUICK GUIDE TO IDEA GENERATION METHODS

	STARTING POINT		RESULT			SKILL LEVEL	
	Generates new ideas	Advances existing ideas	Creates a large idea pool	Generates tangible ideas	Generates out-of-the-box ideas	Requires a high level of creative team maturity	Requires an Idea Agent with some experience
RANDOM WORD ASSOCIATION	✖	✖	✖		✖		✖
FORCED COMBINATIONS		✖		✖			✖
NEGATIVE IDEA GENERATION	✖	✖		✖	✖	✖	✖
WHAT IF?	✖	✖	✖		✖		✖
THE DREAM TRIP	✖	✖		✖	✖	✖	✖
DAY PARTING	✖	✖	✖	✖	✖		
VISUAL CONFETTI	✖	✖	✖		✖		
HEADLINE MANIA		✖		✖			
6-3-5	✖		✖				
FISHING STORIES		✖		✖	✖	✖	✖
IDEA PROPPING	✖	✖	✖	✖	✖		
MERLIN		✖			✖	✖	✖
YOUR CREATIVE IDOL	✖	✖		✖			✖
THE RELAY BATON	✖			✖			
ZOOMING OUT		✖		✖			✖
THE LOTUS BLOSSOM	✖	✖		✖			
TRENDSTORMING	✖	✖	✖	✖			✖
UPSIDE DOWN		✖			✖	✖	✖
TRIZ	✖			✖			✖

D. MODEL SCHEDULE FOR A DAY-LONG WORKSHOP

This is a condensed version of an idea process. A workshop is often just one small piece of the development process puzzle. However, this model schedule will provide a solid framework in which to create your own version.

TIME	METHOD	PURPOSE	MATERIALS
08.30	**INTRODUCTION** – Process background (Project owner). – Introduction to the day and the process (Idea Agent).	Understanding of the root cause and the process goal and purpose.	Projector. Background materials.
08.45	**IDEA GENERATION METHOD: RANDOM WORD ASSOCIATION** – Explain the method's goal and purpose. 08.50 Generate a word association chain. 08.55 Bounce ideas off the focus area – the Idea Agent coaches and controls the pace! 09.15 Each team member clarifies their ideas so that others can read and understand them. 09.20 Collect the ideas in the idea pool.	Freethinking, relaxed team idea generation and a nice energetic start. A large number of wild ideas.	Post-its. Pens. Idea pool.
09.20	**SHORT ENERGIZER**		
09.30	**IDEA GENERATION METHOD: IDEA PROPPING** – Explain the method's goal and purpose. 09.35 Ask team members to write down ideas related to or inspired by the prop they can see in front of them, and place their ideas around it. Change positions every three minutes to a new prop and generate new ideas from the new prop, or develop the existing ideas around it. Idea generation should take place individually and in silence, except possibly for some background music. 10.00 Go to the prop that inspires you most and create new ideas together with any other idea originators you find there. 10.10 Collect the ideas in the idea pool.	Reflective, individual idea generation with physical stimuli. A large number of tangible ideas.	Props. Paper. Pens. Music and CD player.
10.20	**COFFEE BREAK**		
10.30	**IDEA GENERATION METHOD: NEGATIVE IDEA GENERATION** – Explain the method's goal and purpose. 10.35 Discuss as a team a possible inversion for the positive focus area. 10.40 Generate ideas that solve the negative focus area. 11.00 Invert the negative ideas to positive equivalents that solve the positive focus area. 11.25 Collect the ideas in the idea pool.	Light-hearted idea generation that encourages a high energy level in the team. Logical ideas.	Templates for negative idea generation. Paper. Pens.
11.30	**LUNCH**		
12.30	**ASSEMBLY** – Briefly reflect on the morning's activities. – Let team members write down any ideas that may have been generated during lunch. – Briefly describe the afternoon's process.	Assessing the climate in the team and explaining the process direction to them.	
12.45	**IDEA DEVELOPMENT METHOD: THE CLUSTER METHOD** – Explain the method's goal and purpose. 12.50 Cluster the ideas into relevant subject areas. 13.00 When all the ideas have been organized, take one step backwards and grade each cluster 1–10.	Creating an overview of the idea content.	Whiteboard or paper on the wall on which to organize the ideas.

13.15	**SCREENING METHOD: NINE-FIELD MATRIX** – Create a matrix with these axes: Realizable tomorrow/ Realizable/ Not realizable and Revolutionary idea/New idea/Old hat. – Let your team place ideas in the matrix.	Creating an idea overview and carrying out an initial assessment of their quality.	Whiteboard or wallpaper to sketch on.
13.45	**ENRICHMENT** – Explain the graphic enrichment tool. 13.50 Each team member takes an idea from the top right matrix field where the best ideas are placed and begins the enrichment process. 14.20 The team splits into pairs giving feedback and developing each other's ideas. Refreshments are served.	Evolving vague ideas into tangible concepts that are clearly described.	Graphic concept sheets. Pens.
14.45	**PITCHING IDEAS** – Put all the ideas up on the wall. – Ask each team member to present their idea in one minute at most. – Each presentation ends with a round of applause.	Quickly reviewing results. Feeling a sense of achievement.	
15.15	**REFLECTION AND CONCLUSION** – What spontaneous reflections are there from the day's work? – What's the next step in the process? What will happen to the ideas? (Project owner)	Collecting impressions and understanding the next step in the project.	

E. LAZY MAN'S GUIDE TO CREATIVE PROCESS DESIGN

Successful creative processes involve careful planning and interplay between various parameters. The purpose of this comprehensive lazy guide is to assist you with your process planning during creative process design. Use the guide for support and to create an effective process.

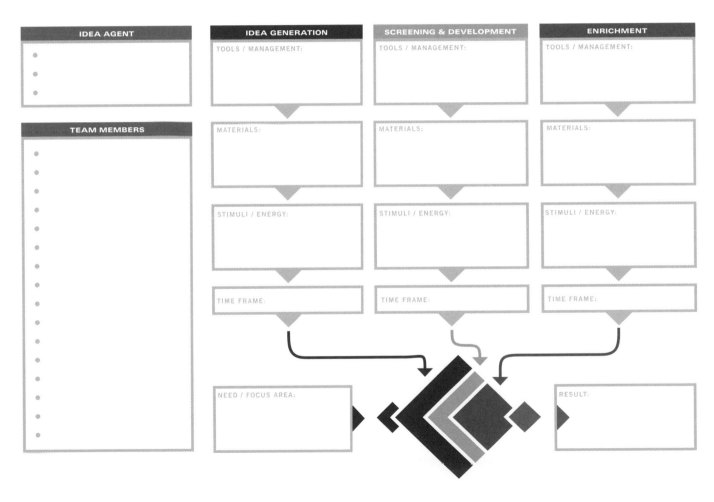

If you would like to subscribe to Idélaboratoriet's creativity and innovation
newsletter, *Serious Innovation*,
please mail to:
news@idelaboratoriet.com

If you want to contact Idélaboratoriet or obtain more information about our
organization and its services, please log on to:
www.idelaboratoriet.com
or follow us at twitter:
@idelaboratoriet